An Introduction
to the Philosophy
of Science

An Introduction to the Philosophy of Science

RUDOLPH CARNAP

EDITED BY
MARTIN GARDNER

DOVER PUBLICATIONS, INC.
New York

Bibliographical Note

This Dover edition, first published in 1995, is an unabridged, corrected republication of the 1974 edition of the work originally published by Basic Books, Inc., New York, 1966, under the title *Philosophical Foundations of Physics: An Introduction to the Philosophy of Science*. For the Dover edition the editor has provided a number of corrections as well as a new Foreword.

Library of Congress Cataloging-in-Publication Data

Carnap, Rudolf, 1891–
    [Philosophical foundations of physics]
    An introduction to the philosophy of science / Rudolph Carnap ; edited by Martin Gardner.
        p.   cm.
    Originally published: Philosophical foundations of physics, New York : Basic Books, 1974.
    Includes bibliographical references and index.
    ISBN 0-486-28318-6 (pbk.)
    1. Physics—Philosophy. I. Gardner, Martin, 1914– .
II. Title.
QC6.C33   1994
530'.01—dc20
                                               94-39458
                                                     CIP

Manufactured in the United States of America
Dover Publications, Inc., 31 East 2nd Street, Mineola, N.Y. 11501

Foreword to the Dover Edition

When I was a freshman at the University of Chicago in 1932, I intended to transfer after two years to the California Institute of Technology to become a physicist. For better or worse, I got sidetracked into philosophy for my bachelor's degree. After one year of graduate work on a scholarship, I decided not to continue for the master's degree but to become a writer instead. I had a job in the university's press-relations office when I enlisted in the Navy.

Back in Chicago after four years of service as a yeoman, I used the G.I. bill in the fall of 1946 to take a seminar with Rudolf Carnap. Titled "Concepts, Theories and Methods in the Physical Sciences," it was the most exciting class I ever attended. It led me into a lifelong interest in the philosophy of science.

After each session I typed the notes I had taken and put them in a looseleaf binder, to which I added an index. Looking over the introductory page in the binder, I find a record reminding me that, during the hour prior to Carnap's class, the same room was used for a course about the "Great Books." (The university was then in its notorious Robert Hutchins–Mortimer Adler phase, which stressed the classics of the Western world.) This often left the blackboard covered with diagrams explaining some aspect of Plato's or Aristotle's metaphysics. Carnap never looked at these diagrams while he erased them. I have likened this sweep of Carnap's arm across the blackboard to his erasure of stale and meaningless metaphysics. One day, before Carnap arrived, a student did the erasing, explaining that it was "so as not to worry Carnap."

Carnap's carefully constructed sentences were delivered slowly, in a low, rich, pleasing voice, giving the impression that he was struggling to simplify ideas too complex for us to fully understand. A favorite phrase, after describing a difficult task, was, "Now how could we do that?" Other often-used phrases included "and we know this for the following reasons," "blurs the distinction," "lacks cognitive content," and "prescientific thinking."

Carnap opened each session with a summary of what he had said at the previous meeting, followed by a period of questions. The philosopher most often cited was Carnap's good friend Hans Reichenbach, with Carl Hempel running second.

I was surprised that Carnap seldom mentioned Bertrand Russell, although I knew he owed Russell a great debt. Later I attended a seminar given by Russell on the Chicago campus. Carnap was in the audience and asking questions. Much of their give and take was beyond me, though I recall Russell saying at one point, "But realism is not a dirty word." They had been arguing over whether it is desirable for a philosopher to assume the reality of an external world as something ontologically certain, or whether realism is no more than the most convenient, indeed indispensable, language for science. Carnap liked to call it the "thing language." Russell soon turned this into a question of whether Carnap's wife was truly "out there" or should be regarded merely as a useful construction within Carnap's experience.

I have written elsewhere about what occurred the next day. I was in the university's post office, talking to philosopher Charles Hartshorne, when Carnap strode in. To my eternal embarrassment, Hartshorne said to Carnap, "Mr. Gardner has been telling me that during Russell's seminar yesterday he tried to persuade you that your wife existed, but you wouldn't admit it."

Carnap glowered at me and said, "But that wasn't the point at all."

Many years later, when Carnap repeated his seminar at the University of California at Los Angeles, I wrote to propose a book. The plan was for someone—who would turn out to be Carnap's wife Ina—to attend the seminar and tape-record each session. She would then type out everything he said, including questions and answers, and send me the pages after each typing. I would edit the material into a coherent volume, working questions and replies into the text as best I could. By then I had started my writing career, and Carnap was familiar with some of my work. He liked the proposal, and the result was the book you now hold. Every idea in the book is Carnap's. Only the phrasing and arrangements are mine. The time I spent working on this book, as copy went back and forth between me and

Carnap for corrections and clarifications, was one of the happiest periods of my life.

Although I never knew Carnap personally and never met his wife, I have vivid and fond memories of his class. He was a teacher who always did his best to make a question, no matter how stupid, seem significant, and to extract from it a meaningful comment. His lectures were extemporaneous, though based on notes he carried on file cards.

It was during this course that Carnap shocked us all by revealing that his friend and former associate Moritz Schlick had just been murdered by a psychotic student. There were, however, moments of comedy as well. I remember one confusing interchange with a woman mathematics teacher before it was discovered that she was using the word "pear," the fruit: Carnap had taken it to be "pair"—or maybe it was the other way around.

Basic Books published our book in 1966 under Carnap's preferred title *The Philosophical Foundations of Physics: An Introduction to the Philosophy of Science.* After Carnap's death in 1970, Basic reissued the book in paper covers, and the original subtitle became the new title, which has been retained in the Dover edition. Many corrections for the Basic edition were generously supplied by Carnap's friend Carl G. Hempel. I had asked Hempel to write a foreword, but he declined because he considered it inappropriate to write a foreword to a book by a person so much more eminent than he.

The book received good reviews, and was adopted in the classroom by Wesley Salmon and a few other noted philosophers of science here and abroad. Translations were published in Germany, France, Italy, Japan, and Argentina. It should be emphasized that this is the only book by Carnap on a level sufficiently nontechnical to be understood by readers with no expertise in mathematics, physics, or logic.

For the Dover edition I am indebted to Dover president Hayward Cirker and to editor John Grafton for recognizing the book's merit, as well as to retired philosopher of science Arthur J. Benson for dozens of corrections that have been made throughout. It was Professor Benson who had compiled the bibliography for *The Philosophy of Rudolph Carnap*, edited by Paul Arthur Schilpp as the eleventh volume of the distinguished Library of Living Philosophers series.

I also wish to thank Carnap's daughter Hanna Carnap Thost for allowing Dover to reissue this book and for putting me in touch with Dr. Benson.

It was a great privilege to have attended Carnap's seminar and to have been given the honor of editing his book. Although Carnap's reputation is not as high now as it was then, I have no doubt that it will steadily rise again. More and more younger philosophers of science will surely discover the greatness of his contributions and his influence, and how right he was in his notable quarrels (in my opinion largely verbal quibbles) with Karl Popper and Willard Van Orman Quine.

MARTIN GARDNER
Hendersonville, N.C.

Foreword to the Basic Books paperback edition, 1974

One of the most memorable privileges of my life was to have attended Rudolf Carnap's seminar on "philosophical foundations of physics" when he was at the University of Chicago. It was an even greater privilege, many years later, to be allowed to shape those seminar lectures (after Carnap had repeated them at the University of California) into the present volume. Although not exactly an elementary or "popular" book, it is certainly much less technical than any of Carnap's other works. In my opinion it is the best first introduction to the views of one of this century's great creative philosophers, as well as one of the clearest and soundest of modern introductions to the philosophy of science.

The book originally bore the title Carnap had often used for his seminar, followed by the subtitle: "An introduction to the philosophy of science." Wesley C. Salmon is mainly responsible for this edition's switch of the two phrases. Salmon had given the book a splendid review (*Science,* March 10, 1967), and for several years had used the volume for assigned reading in his classes. Two years ago he made two suggestions: first, that the book be reprinted in a paperback edition students could afford; second, that its formidable title, which conveys the false impression of a highly technical work, be changed. Both proposals have now been adopted.

Aside from a few trivial corrections, the only important textual changes are on pages 255 and 256. In response to a friendly letter from Grover Maxwell, Carnap agreed (shortly before his death in 1970) that his all-too-brief comments on the conflict between instrumentalism and realism, with respect to the nature of scientific theory, be clarified. With this in mind, he made certain alterations on the two pages, and added a new footnote referring to a 1950 paper which gives his views in more detail.

I have resisted a temptation to update the Bibliography, preferring to leave it as Carnap wanted it in 1966 rather than make my own selections from the many excellent books that have appeared since.

MARTIN GARDNER

Preface

This book grew out of a seminar that I have given many times, with varying content and form. It was called "Philosophical foundations of physics" or "Concepts, theories, and methods of the physical sciences." Although the content often changed, the general philosophical point of view remained constant; the course emphasized the logical analysis of the concepts, statements, and theories of science, rather than metaphysical speculation.

The idea of presenting the substance of my (rather informal) seminar talks in a book was suggested by Martin Gardner, who had attended my course in 1946 at the University of Chicago. He inquired in 1958 whether a typescript of the seminar existed or could be made; if so, he offered to edit it for publication. I have never had typescripts of my lectures or seminar talks, and I was not willing to take the time to write one. It just happened that this course was announced for the next semester, Fall 1958, at the University of California at Los Angeles. It was suggested that my talks and the discussions be recorded. Conscious of the enormous distance between the spoken word and a formulation suitable for publication, I was first rather skeptical about the plan. But my friends urged me to do it, because not many of my views on problems in the philosophy of science had been published. The decisive encouragement came from my wife, who volunteered to record the whole semester course on tape and transcribe it. She did this and also gave me invaluable help in the later phases of the working process. The book owes much to her; but she did not live to see it published.

A corrected version of the transcript was sent to Martin Gardner. Then he began his difficult task, which he carried out with great skill and sensitivity. He not only smoothed out the style, but found ways of making the reading easier by rearranging some of the topics and by improving examples or contributing new ones. The chapters went back and forth several times. Now and then, I made extensive changes or additions or suggested to Gardner that he make them. Although the seminar was for advanced graduate students in philosophy who were familiar with symbolic logic and had some knowledge of college mathematics and physics, we decided to make the book accessible to a wider circle of readers. The number of logical, mathematical, and physical formulas was considerably reduced, and the remaining ones were explained wherever it seemed advisable.

No attempt is made in this book to give a systematic treatment of all the important problems in the philosophical foundations of physics. In my seminar—therefore also in the book—I have preferred to restrict myself to a small number of fundamental problems (as indicated by the headings of the six parts) and to discuss them more thoroughly, instead of including a cursory discussion of many other subjects. Most of the topics dealt with in this book (except for Part III, on geometry, and Chapter 30, on quantum physics) are relevant to all branches of science, including the biological sciences, psychology, and the social sciences. I believe, therefore, that this book may also serve as a general introduction to the philosophy of science.

My first thanks go to my faithful and efficient collaborator, Martin Gardner. I am grateful for his excellent work and also for his inexhaustible patience when I made long delays in returning some chapters or asked for still more changes.

My friends Herbert Feigl and Carl G. Hempel I wish to thank for suggestive ideas they presented in conversations through many years and especially for their helpful comments on parts of the manuscript. I thank Abner Shimony for generous expert help on questions concerning quantum mechanics. And, further, I am grateful to many friends and colleagues for their stimulating influence and to my students who attended one or another version of this seminar and whose questions and comments prompted some of the discussions in this book.

I acknowledge with thanks the kind permission of Yale University Press for extensive quotations from Kurt Riezler's book, *Physics and Reality* (1940).

RUDOLF CARNAP

University of California at Los Angeles

February 1966

Contents

Part I

LAWS,
EXPLANATION,
AND PROBABILITY

CHAPTER 1

The Value of Laws:
Explanation
and Prediction

THE OBSERVATIONS we make in every-
day life as well as the more systematic observations of science reveal
certain repetitions or regularities in the world. Day always follows night;
the seasons repeat themselves in the same order; fire always feels hot;
objects fall when we drop them; and so on. The laws of science are nothing more than statements expressing these regularities as precisely as
possible.

If a certain regularity is observed at all times and all places, without exception, then the regularity is expressed in the form of a "universal law". An example from daily life is, "All ice is cold." This statement asserts that any piece of ice—at any place in the universe, at any
time, past, present, or future—is (was, or will be) cold. Not all laws of
science are universal. Instead of asserting that a regularity occurs in *all*
cases, some laws assert that it occurs in only a certain percentage of
cases. If the percentage is specified or if in some other way a quantitative
statement is made about the relation of one event to another, then the
statement is called a "statistical law". For example: "Ripe apples are
usually red", or "Approximately half the children born each year are

3

boys." Both types of law—universal and statistical—are needed in science. The universal laws are logically simpler, and for this reason we shall consider them first. In the early part of this discussion "laws" will usually mean universal laws.

Universal laws are expressed in the logical form of what is called in formal logic a "universal conditional statement". (In this book, we shall occasionally make use of symbolic logic, but only in a very elementary way.) For example, let us consider a law of the simplest possible type. It asserts that, whatever x may be, if x is P, then x is also Q. This is written symbolically as follows:

$$(x)(Px \supset Qx).$$

The expression "(x)" on the left is called a "universal quantifier." It tells us that the statement refers to *all* cases of x, rather than to just a certain percentage of cases. "Px" says that x is P, and "Qx" says that x is Q. The symbol "$\supset$" is a connective. It links the term on its left to the term on its right. In English, it corresponds roughly to the assertion, "If . . . then . . ."

If "x" stands for any material body, then the law states that, for any material body x, if x has the property P, it also has the property Q. For instance, in physics we might say: "For every body x, if that body is heated, that body will expand." This is the law of thermal expansion in its simplest, nonquantitative form. In physics, of course, one tries to obtain quantitative laws and to qualify them so as to exclude exceptions; but, if we forget about such refinements, then this universal conditional statement is the basic logical form of all universal laws. Sometimes we may say that, not only does Qx hold whenever Px holds, but the reverse is also true; whenever Qx holds, Px holds also. Logicians call this a biconditional statement—a statement that is conditional in both directions. But of course this does not contradict the fact that in all universal laws we deal with universal conditionals, because a biconditional may be regarded as the conjunction of two conditionals.

Not all statements made by scientists have this logical form. A scientist may say: "Yesterday in Brazil, Professor Smith discovered a new species of butterfly." This is not the statement of a law. It speaks about a specified single time and place; it states that something happened at that time and place. Because statements such as this are about single facts, they are called "singular" statements. Of course, all our knowledge has its origin in singular statements—the particular observations of particular individuals. One of the big, perplexing questions in the philosophy of science is how we are able to go from such singular statements to the assertion of universal laws.

When statements by scientists are made in the ordinary word language, rather than in the more precise language of symbolic logic, we must be extremely careful not to confuse singular with universal statements. If a zoologist writes in a textbook, "The elephant is an excellent swimmer", he does not mean that a certain elephant, which he observed a year ago in a zoo, is an excellent swimmer. When he says "the elephant", he is using "the" in the Aristotelian sense; it refers to the entire class of elephants. All European languages have inherited from the Greek (and perhaps also from other languages) this manner of speaking in a singular way when actually a class or type is meant. The Greeks said, "Man is a rational animal." They meant, of course, all men, not a particular man. In a similar way, we say "the elephant" when we mean all elephants or "tuberculosis is characterized by the following symptoms . . ." when we mean, not a singular case of tuberculosis, but all instances.

It is unfortunate that our language has this ambiguity, because it is a source of much misunderstanding. Scientists often refer to universal statements—or rather to what is expressed by such statements—as "facts". They forget that the word "fact" was originally applied (and we shall apply it exclusively in this sense) to singular, particular occurrences. If a scientist is asked about the law of thermal expansion, he may say: "Oh, thermal expansion. That is one of the familiar, basic facts of physics." In a similar way, he may speak of the fact that heat is generated by an electric current, the fact that magnetism is produced by electricity, and so on. These are sometimes considered familiar "facts" of physics. To avoid misunderstandings, we prefer not to call such statements "facts". Facts are particular events. "This morning in the laboratory, I sent an electric current through a wire coil with an iron body inside it, and I found that the iron body became magnetic." That is a fact unless, of course, I deceived myself in some way. However, if I was sober, if it was not too foggy in the room, and if no one has tinkered secretly with the apparatus to play a joke on me, then I may state as a factual observation that this morning that sequence of events occurred.

When we use the word "fact", we will mean it in the singular sense in order to distinguish it clearly from universal statements. Such universal statements will be called "laws" even when they are as elementary as the law of thermal expansion or, still more elementary, the statement,

"All ravens are black." I do not know whether this statement is true, but, assuming its truth, we will call such a statement a law of zoology. Zoologists may speak informally of such "facts" as "the raven is black" or "the octopus has eight arms", but, in our more precise terminology, statements of this sort will be called "laws".

Later we shall distinguish between two kinds of law—empirical and theoretical. Laws of the simple kind that I have just mentioned are sometimes called "empirical generalizations" or "empirical laws". They are simple because they speak of properties, like the color black or the magnetic properties of a piece of iron, that can be directly observed. The law of thermal expansion, for example, is a generalization based on many direct observations of bodies that expand when heated. In contrast, theoretical, nonobservable concepts, such as elementary particles and electromagnetic fields, must be dealt with by theoretical laws. We will discuss all this later. I mention it here because otherwise you might think that the examples I have given do not cover the kind of laws you have perhaps learned in theoretical physics.

To summarize, science begins with direct observations of single facts. Nothing else is observable. Certainly a regularity is not directly observable. It is only when many observations are compared with one another that regularities are discovered. These regularities are expressed by statements called "laws".

What good are such laws? What purposes do they serve in science and everyday life? The answer is twofold: they are used to *explain* facts already known, and they are used to *predict* facts not yet known.

First, let us see how laws of science are used for explanation. No explanation—that is, nothing that deserves the honorific title of "explanation"—can be given without referring to at least one law. (In simple cases, there is only one law, but in more complicated cases a set of many laws may be involved.) It is important to emphasize this point, because philosophers have often maintained that they could explain certain facts in history, nature, or human life in some other way. They usually do this by specifying some type of agent or force that is made responsible for the occurrence to be explained.

In everyday life, this is, of course, a familiar form of explanation. Someone asks: "How is it that my watch, which I left here on the table before I left the room, is no longer here?" You reply: "I saw Jones come into the room and take it." This is your explanation of the watch's disappearance. Perhaps it is not considered a sufficient explanation. Why did Jones take the watch? Did he intend to steal it or just to borrow it? Perhaps he took it under the mistaken impression that it was his own. The first question, "What happened to the watch?", was answered by a statement of fact: Jones took it. The second question, "Why did Jones take it?", may be answered by another fact: he borrowed it for a moment. It seems, therefore, that we do not need laws at all. We ask for an explanation of one fact, and we are given a second fact. We ask for an explanation of the second fact, and we are given a third. Demands for further explanations may bring out still other facts. Why, then, is it necessary to refer to a law in order to give an adequate explanation of a fact?

The answer is that fact explanations are really law explanations in disguise. When we examine them more carefully, we find them to be abbreviated, incomplete statements that tacitly assume certain laws, but laws so familiar that it is unnecessary to express them. In the watch illustration, the first answer, "Jones took it", would not be considered a satisfactory explanation if we did not assume the universal law: whenever someone takes a watch from a table, the watch is no longer on the table. The second answer, "Jones borrowed it", is an explanation because we take for granted the general law: if someone borrows a watch to use elsewhere, he takes the watch and carries it away.

Consider one more example. We ask little Tommy why he is crying, and he answers with another fact: "Jimmy hit me on the nose." Why do we consider this a sufficient explanation? Because we know that a blow on the nose causes pain and that, when children feel pain, they cry. These are general psychological laws. They are so well known that they are assumed even by Tommy when he tells us why he is crying. If we were dealing with, say, a Martian child and knew very little about Martian psychological laws, then a simple statement of fact might not be considered an adequate explanation of the child's behavior. Unless facts can be connected with other facts by means of at least one law, explicitly stated or tacitly understood, they do not provide explanations.

The general schema involved in all explanation of the deductive variety can be expressed symbolically as follows:

1. $(x)(Px \supset Qx)$
2. Pa
3. Qa

The first statement is the universal law that applies to any object x. The second statement asserts that a particular object a has the property

P. These two statements taken together enable us to derive logically the third statement: object *a* has the property *Q*.

In science, as in everyday life, the universal law is not always explicitly stated. If you ask a physicist: "Why is it that this iron rod, which a moment ago fitted exactly into the apparatus, is now a trifle too long to fit?", he may reply by saying: "While you were out of the room, I heated the rod." He assumes, of course, that you know the law of thermal expansion; otherwise, in order to be understood, he would have added, "and, whenever a body is heated, it expands". The general law is essential to his explanation. If you know the law, however, and he knows that you know it, he may not feel it necessary to state the law. For this reason, explanations, especially in everyday life where common-sense laws are taken for granted, often seem quite different from the schema I have given.

At times, in giving an explanation, the only known laws that apply are statistical rather than universal. In such cases, we must be content with a statistical explanation. For example, we may know that a certain kind of mushroom is slightly poisonous and causes certain symptoms of illness in 90 per cent of those who eat it. If a doctor finds these symptoms when he examines a patient and the patient informs the doctor that yesterday he ate this particular kind of mushroom, the doctor will consider this an explanation of the symptoms even though the law involved is only a statistical one. And it is, indeed, an explanation.

Even when a statistical law provides only an extremely weak explanation, it is still an explanation. For instance, a statistical medical law may state that 5 per cent of the people who eat a certain food will develop a certain symptom. If a doctor cites this as his explanation to a patient who has the symptom, the patient may not be satisfied. "Why", he asks, "am I one of the 5 per cent?" In some cases, the doctor may be able to provide further explanations. He may test the patient for allergies and find that he is allergic to this particular food. "If I had known this", he tells the patient, "I would have warned you against this food. We know that, when people who have such an allergy eat this food, 97 per cent of them will develop symptoms such as yours." That may satisfy the patient as a stronger explanation. Whether strong or weak, these are genuine explanations. In the absence of known universal laws, statistical explanations are often the only type available.

In the example just given, the statistical laws are the best that can be stated, because there is not sufficient medical knowledge to warrant stating a universal law. Statistical laws in economics and other fields of social science are due to a similar ignorance. Our limited knowledge of psychological laws, of the underlying physiological laws, and of how those may in turn rest on physical laws makes it necessary to formulate the laws of social science in statistical terms. In quantum theory, however, we meet with statistical laws that may not be the result of ignorance; they may express the basic structure of the world. Heisenberg's famous principle of uncertainty is the best-known example. Many physicists believe that all the laws of physics rest ultimately on fundamental laws that are statistical. If this is the case, we shall have to be content with explanations based on statistical laws.

What about the elementary laws of logic that are involved in all explanations? Do they ever serve as the universal laws on which scientific explanation rests? No, they do not. The reason is that they are laws of an entirely different sort. It is true that the laws of logic and pure mathematics (not physical geometry, which is something else) are universal, but they tell us nothing whatever about the world. They merely state relations that hold between certain concepts, not because the world has such and such a structure, but only because those concepts are defined in certain ways.

Here are two examples of simple logical laws:

1. If *p* and *q*, then *p*.
2. If *p*, then *p* or *q*.

Those statements cannot be contested because their truth is based on the meanings of the terms involved. The first law merely states that, if we assume the truth of statements *p* and *q*, then we must assume that statement *p* is true. The law follows from the way in which "and" and "if . . . then" are used. The second law asserts that, if we assume the truth of *p*, we must assume that either *p* or *q* is true. Stated in words, the law is ambiguous because the English "or" does not distinguish between an inclusive meaning (either or both) and the exclusive meaning (either but not both). To make the law precise, we express it symbolically by writing:

$$p \supset (p \lor q)$$

The symbol "$\lor$" is understood as "or" in the inclusive sense. Its meaning can be given more formally by writing out its truth table. We do this by listing all possible combinations of truth values (truth or falsity) for the two terms connected by the symbol, then specifying which combinations are permitted by the symbol and which are not.

The four possible combinations of values are:

	p	q
1.	true	true
2.	true	false
3.	false	true
4.	false	false

The symbol "$\vee$" is defined by the rule that "$p \vee q$" is true in the first three cases and false in the fourth case. The symbol "$\supset$", which translates roughly into English as "if . . . then", is defined by saying that $p \supset q$ is true in the first, third, and fourth cases, and false in the second. Once we understand the definition of each term in a logical law, we see clearly that the law must be true in a way that is wholly independent of the nature of the world. It is a necessary truth, a truth that holds, as philosophers sometimes put it, in all possible worlds.

This is true of the laws of mathematics as well as those of logic. When we have precisely specified the meanings of "1", "3", "4", "+", and "=", the truth of the law "$1 + 3 = 4$" follows directly from these meanings. This is the case even in the more abstract areas of pure mathematics. A structure is called a "group", for example, if it fulfills certain axioms that define a group. Three-dimensional Euclidean space can be defined algebraically as a set of ordered triples of real numbers that fulfill certain basic conditions. But all this has nothing to do with the nature of the outside world. There is no possible world in which the laws of group theory and the abstract geometry of Euclidean 3-space would not hold, because these laws are dependent only on the meanings of the terms involved, and not on the structure of the actual world in which we happen to be.

The actual world is a world that is constantly changing. Even the most fundamental laws of physics may, for all we can be sure, vary slightly from century to century. What we believe to be a physical constant with a fixed value may be subject to vast cyclic changes that we have not yet observed. But such changes, no matter how drastic, would never destroy the truth of a single logical or arithmetical law.

It sounds very dramatic, perhaps comforting, to say that here at last we have actually found certainty. It is true that we have obtained certainty, but we have paid for it a very high price. The price is that statements of logic and mathematics do not tell us anything about the world. We can be sure that three plus one is four; but, because this holds in any possible world, it can tell us nothing whatever about the world we inhabit.

What do we mean by "possible world"? Simply a world that can be described without contradiction. It includes fairy-tale worlds and dream worlds of the most fantastic kind, provided that they are described in logically consistent terms. For example, you may say: "I have in mind a world in which there are exactly one thousand events, no more, no less. The first event is the appearance of a red triangle. The second is the appearance of a green square. However, since the first event was blue and not red . . .". At this point, I interrupt. "But a moment ago you said that the first event is red. Now you say that it is blue. I do not understand you." Perhaps I have recorded your remarks on tape. I play back the tape to convince you that you have stated a contradiction. If you persist in your description of this world, including the two contradictory assertions, I would have to insist that you are not describing anything that can be called a possible world.

On the other hand, you may describe a possible world as follows: "There is a man. He shrinks in size, becoming smaller and smaller. Suddenly he turns into a bird. Then the bird becomes a thousand birds. These birds fly into the sky, and the clouds converse with one another about what happened." All this is a possible world. Fantastic, yes; contradictory, no.

We might say that possible worlds are conceivable worlds, but I try to avoid the term "conceivable" because it is sometimes used in the more restricted sense of "what can be imagined by a human being". Many possible worlds can be described but not imagined. We might, for example, discuss a continuum in which all points determined by rational coordinates are red and all points determined by irrational coordinates are blue. If we admit the possibility of ascribing colors to points, this is a noncontradictory world. It is conceivable in the wider sense; that is, it can be assumed without contradiction. It is not conceivable in the psychological sense. No one can imagine even an uncolored continuum of points. We can imagine only a crude model of a continuum—a model consisting of very tightly packed points. Possible worlds are worlds that are conceivable in the wider sense. They are worlds that can be described without logical contradiction.

The laws of logic and pure mathematics, by their very nature, cannot be used as a basis for scientific explanation because they tell us nothing that distinguishes the actual world from some other possible world.

When we ask for the explanation of a fact, a particular observation in the actual world, we must make use of *empirical* laws. They do not possess the certainty of logical and mathematical laws, but they do tell us something about the structure of the world.

In the nineteenth century, certain Germanic physicists, such as Gustav Kirchhoff and Ernst Mach, said that science should not ask "Why?" but "How?" They meant that science should not look for unknown metaphysical agents that are responsible for certain events, but should only describe such events in terms of laws. This prohibition against asking "Why?" must be understood in its historical setting. The background was the German philosophical atmosphere of the time, which was dominated by idealism in the tradition of Fichte, Schelling, and Hegel. These men felt that a description of how the world behaved was not enough. They wanted a fuller understanding, which they believed could be obtained only by finding metaphysical causes that were behind phenomena and not accessible to scientific method. Physicists reacted to this point of view by saying: "Leave us alone with your why-questions. There is no answer beyond that given by the empirical laws." They objected to why-questions because they were usually metaphysical questions.

Today the philosophical atmosphere has changed. In Germany there are a few philosophers still working in the idealist tradition, but in England and the United States it has practically disappeared. As a result, we are no longer worried by why-questions. We do not have to say, "Don't ask why", because now, when someone asks why, we assume that he means it in a scientific, nonmetaphysical sense. He is simply asking us to explain something by placing it in a framework of empirical laws.

When I was young and part of the Vienna Circle, some of my early publications were written as a reaction to the philosophical climate of German idealism. As a consequence, these publications and those by others in the Vienna Circle were filled with prohibitory statements similar to the one I have just discussed. These prohibitions must be understood in reference to the historical situation in which we found ourselves. Today, especially in the United States, we seldom make such prohibitions. The kind of opponents we have here are of a different nature, and the nature of one's opponents often determines the way in which one's views are expressed.

When we say that, for the explanation of a given fact, the use of a scientific law is indispensable, what we wish to exclude especially is the view that metaphysical agents must be found before a fact can be adequately explained. In prescientific ages, this was, of course, the kind of explanation usually given. At one time, the world was thought to be inhabited by spirits or demons who are not directly observable but who *act* to cause the rain to fall, the river to flow, the lightning to flash. In whatever one saw happening, there was something—or, rather, *somebody*—responsible for the event. This is psychologically understandable. If a man does something to me that I do not like, it is natural for me to make him responsible for it and to get angry and hit back at him. If a cloud pours water over me, I cannot hit back at the cloud, but I can find an outlet for my anger if I make the cloud, or some invisible demon behind the cloud, responsible for the rainfall. I can shout curses at this demon, shake my fist at him. My anger is relieved. I feel better. It is easy to understand how members of prescientific societies found psychological satisfaction in imagining agents behind the phenomena of nature.

In time, as we know, societies abandoned their mythologies, but sometimes scientists replace the spirits with agents that are really not much different. The German philosopher Hans Driesch, who died in 1941, wrote many books on the philosophy of science. He was originally a prominent biologist, famed for his work on certain organismic responses, including regeneration in sea urchins. He cut off parts of their bodies and observed in which stages of their growth and under what conditions they were able to grow new parts. His scientific work was important and excellent. But Driesch was also interested in philosophical questions, especially those dealing with the foundations of biology, so eventually he became a professor of philosophy. In philosophy also he did some excellent work, but there was one aspect of his philosophy that I and my friends in the Vienna Circle did not regard so highly. It was his way of *explaining* such biological processes as regeneration and reproduction.

At the time Driesch did his biological work, it was thought that many characteristics of living things could not be found elsewhere. (Today it is seen more clearly that there is a continuum connecting the organic and inorganic worlds.) He wanted to explain these unique organismic features, so he postulated what he called an "entelechy". This term had been introduced by Aristotle, who had his own meaning for it, but we need not discuss that meaning here. Driesch said, in effect: "The entelechy is a certain specific force that causes living things to behave in the way they do. But you must not think of it as a *physical* force such as gravity or magnetism. Oh, no, nothing like that."

The entelechies of organisms, Driesch maintained, are of various kinds, depending on the organism's stage of evolution. In primitive, single-celled organisms, the entelechy is rather simple. As we go up the evolutionary scale, through plants, lower animals, higher animals, and finally to man, the entelechy becomes more and more complex. This is revealed by the greater degree to which phenomena are integrated in the higher forms of life. What we call the "mind" of a human body is actually nothing more than a portion of the person's entelechy. The entelechy is much more than the mind, or, at least, more than the conscious mind, because it is responsible for everything that every cell in the body does. If I cut my finger, the cells of the finger form new tissue and bring substances to the cut to kill incoming bacteria. These events are not consciously directed by the mind. They occur in the finger of a one-month-old baby, who has never heard of the laws of physiology. All this, Driesch insisted, is due to the organism's entelechy, of which mind is *one* manifestation. In addition, then, to scientific explanation, Driesch had an elaborate theory of entelechy, which he offered as a *philosophical* explanation of such scientifically unexplained phenomena as the regeneration of parts of sea urchins.

Is this an explanation? I and my friends had some discussions with Driesch about it. I remember one at the International Congress for Philosophy, at Prague, in 1934. Hans Reichenbach and I criticized Driesch's theory, while he and others defended it. In our publications we did not give much space to this criticism because we admired the work Driesch had done in both biology and philosophy. He was quite different from most philosophers in Germany in that he really wanted to develop a scientific philosophy. His entelechy theory, however, seemed to us to lack something.

What it lacked was this: the insight that you cannot give an explanation without also giving a law.

We said to him: "Your entelechy—we do not know what you mean by it. You say it is not a physical force. What is it then?"

"Well", he would reply (I am paraphrasing his words, of course), "you should not be so narrow-minded. When you ask a physicist for an explanation of why this nail suddenly moves toward that bar of iron, he will tell you that the bar of iron is a magnet and that the nail is drawn to it by the force of magnetism. No one has ever seen magnetism. You see only the movement of a little nail toward a bar of iron."

We agreed. "Yes, you are right. Nobody has seen magnetism."

"You see", he continued, "the physicist introduces forces that no one can observe—forces like magnetism and electricity—in order to explain certain phenomena. I wish to do the same. Physical forces are not adequate to explain certain organic phenomena, so I introduce something that is forcelike but is not a physical force because it does not act the way physical forces act. For instance, it is not spatially located. True, it acts on a physical organism, but it acts in respect to the entire organism, not just to certain parts of it. Therefore, you cannot say where it is located. There is no location. It is not a physical force, but it is just as legitimate for me to introduce it as it is for a physicist to introduce the invisible force of magnetism."

Our answer was that a physicist does not explain the movement of the nail toward the bar simply by introducing the word "magnetism". Of course, if you ask him why the nail moves, he may answer first by saying that it is due to magnetism; but if you press him for a fuller explanation, he will give you laws. The laws may not be expressed in quantitative terms, like the Maxwell equations that describe magnetic fields; they may be simple, qualitative laws with no numbers occurring in them. The physicist may say: "All nails containing iron are attracted to the ends of bars that have been magnetized." He may go on to explain the state of being magnetized by giving other nonquantitative laws. He may tell you that iron ore from the town of Magnesia (you may recall that the word "magnetic" derives from the Greek town of Magnesia, where iron ore of this type was first found) possesses this property. He may explain that iron bars become magnetized if they are stroked a certain way by naturally magnetic ores. He may give you other laws about conditions under which certain substances can become magnetized and laws about phenomena associated with magnetism. He may tell you that if you magnetize a needle and suspend it by the middle so that it swings freely, one end will point north. If you have another magnetic needle, you can bring the two north-pointing ends together and observe that they do not attract but repel each other. He may explain that if you heat a magnetized bar of iron, or hammer it, it will lose magnetic strength. All these are qualitative laws that can be expressed in the logical form, "if . . . then . . ." The point I wish to emphasize here is this: it is not sufficient, for purposes of explanation, simply to introduce a new agent by giving it a new name. You must also give laws.

Driesch did not give laws. He did not specify how the entelechy of an oak tree differs from the entelechy of a goat or giraffe. He did not classify his entelechies. He merely classified organisms and said that each organism had its own entelechy. He did not formulate laws that state under what conditions an entelechy is strengthened or weakened. Of course he described all sorts of organic phenomena and gave general rules for such phenomena. He said that if you cut a limb from a sea urchin in a certain way, the organism will not survive; if you cut it another way, the organism will survive, but only a fragmentary limb will grow back. Cut in still another way and at a certain stage in the sea urchin's growth, it will regenerate a new and complete limb. These statements are all perfectly respectable zoological laws.

"What do you add to these empirical laws", we asked Driesch, "if after giving them you proceed to tell us that all the phenomena covered by those laws are due to the sea urchin's entelechy?"

We believed that nothing was added. Since the notion of an entelechy does not give us new laws, it does not explain more than the general laws already available. It does not help us in the least in making new predictions. For these reasons we cannot say that our scientific knowledge has increased. The concept of entelechy may at first seem to add something to our explanations; but when we examine it more deeply, we see its emptiness. It is a pseudoexplanation.

It can be argued that the concept of entelechy is not useless if it provides biologists with a new orientation, a new method of ordering biological laws. Our answer is that it would indeed be useful if by means of it we could formulate more general laws than could be formulated before. In physics, for example, the concept of energy played such a role. Nineteenth-century physicists theorized that perhaps certain phenomena, such as kinetic and potential energy in mechanics, heat (this was before the discovery that heat is simply the kinetic energy of molecules), the energy of magnetic fields, and so on, might be manifestations of one basic kind of energy. This led to experiments showing that mechanical energy can be transformed into heat and heat into mechanical energy but that the amount of energy remains constant. Thus, energy was a fruitful concept because it led to more general laws, such as the law of the conservation of energy. But Driesch's entelechy was not a fruitful concept in this sense. It did not lead to the discovery of more general biological laws.

In addition to providing *explanations* for observed facts, the laws of science also provide a means for *predicting* new facts not yet observed. The logical schema involved here is exactly the same as the schema underlying explanation. This, you recall, was expressed symbolically:

1. $(x)(Px \supset Qx)$
2. Pa
3. Qa

First we have a universal law: for any object x, if it has the property P, then it also has the property Q. Second, we have a statement saying that object a has the property P. Third, we deduce by elementary logic that object a has the property Q. This schema underlies both explanation and prediction; only the knowledge situation is different. In explanation, the fact Qa is already known. We explain Qa by showing how it can be deduced from statements 1 and 2. In prediction, Qa is a fact *not yet known*. We have a law, and we have the fact Pa. We conclude that Qa must also be a fact, even though it has not yet been observed. For example, I know the law of thermal expansion. I also know that I have heated a certain rod. By applying logic in the way shown in the schema, I infer that if I now measure the rod, I will find that it is longer than it was before.

In most cases, the unknown fact is actually a future event (for example, an astronomer predicts the time of the next eclipse of the sun); that is why I use the term "prediction" for this second use of laws. It need not, however, be prediction in the literal sense. In many cases the unknown fact is simultaneous with the known fact, as is the case in the example of the heated rod. The expansion of the rod occurs simultaneously with the heating. It is only our observation of the expansion that takes place after our observation of the heating.

In other cases, the unknown fact may even be in the past. On the basis of psychological laws, together with certain facts derived from historical documents, a historian infers certain unknown facts of history. An astronomer may infer that an eclipse of the moon must have taken place at a certain date in the past. A geologist may infer from striations on boulders that at one time in the past a region must have been covered by a glacier. I use the term "prediction" for all these examples because in every case we have the same logical schema and the same knowledge situation—a known fact and a known law from which an unknown fact is derived.

In many cases, the law involved may be statistical rather than universal. The prediction will then be only probable. A meteorologist, for instance, deals with a mixture of exact physical laws and various statistical laws. He cannot say that it will rain tomorrow; he can only say that rain is very likely.

This uncertainty is also characteristic of prediction about human behavior. On the basis of knowing certain psychological laws of a statistical nature and certain facts about a person, we can predict with varying degrees of probability how he will behave. Perhaps we ask a psychologist to tell us what effect a certain event will have on our child. He replies: "As I see the situation, your child will probably react in this way. Of course, the laws of psychology are not very exact. It is a young science, and as yet we know very little about its laws. But on the basis of what is known, I think it advisable that you plan to . . .". And so he gives us advice based on the best prediction he can make, with his probabilistic laws, about the future behavior of our child.

When the law is universal, then elementary deductive logic is involved in inferring unknown facts. If the law is statistical, we must use a different logic—the logic of probability. To give a simple example: a law states that 90 per cent of the residents of a certain region have black hair. I know that an individual is a resident of that region, but I do not know the color of his hair. I can infer, however, on the basis of the statistical law, that the probability his hair is black is $\frac{9}{10}$.

Prediction is, of course, as essential to everyday life as it is to science. Even the most trivial acts we perform during the day are based on predictions. You turn a doorknob. You do so because past observations of facts, together with universal laws, lead you to believe that turning the knob will open the door. You may not be conscious of the logical schema involved—no doubt you are thinking about other things—but all such deliberate actions presuppose the schema. There is a knowledge of specific facts, a knowledge of certain observed regularities that can be expressed as universal or statistical laws and provide a basis for the prediction of unknown facts. Prediction is involved in every act of human behavior that involves deliberate choice. Without it, both science and everyday life would be impossible.

CHAPTER 2

Induction and
Statistical Probability

IN CHAPTER 1, we assumed that laws of science were available. We saw how such laws are used, in both science and everyday life, as explanations of known facts and as a means for predicting unknown facts. Let us now ask how we arrive at such laws. On what basis are we justified in believing that a law holds? We know, of course, that all laws are based on the observation of certain regularities. They constitute indirect knowledge, as opposed to direct knowledge of facts. What justifies us in going from the direct observation of facts to a law that expresses certain regularities of nature? This is what in traditional terminology is called "the problem of induction".

Induction is often contrasted with deduction by saying that deduction goes from the general to the specific or singular, whereas induction goes the other way, from the singular to the general. This is a misleading oversimplification. In deduction, there are kinds of inferences other than those from the general to the specific; in induction there are also many kinds of inference. The traditional distinction is also misleading because it suggests that deduction and induction are simply two branches of a single kind of logic. John Stuart Mill's famous work, *A*

System of Logic, contains a lengthy description of what he called "inductive logic" and states various canons of inductive procedure. Today we are more reluctant to use the term "inductive inference". If it is used at all, we must realize that it refers to a kind of inference that differs fundamentally from deduction.

In deductive logic, inference leads from a set of premisses to a conclusion just as certain as the premisses. If you have reason to believe the premisses, you have equally valid reason to believe the conclusion that follows logically from the premisses. If the premisses are true, the conclusion cannot be false. With respect to induction, the situation is entirely different. The truth of an inductive conclusion is never certain. I do not mean only that the conclusion cannot be certain because it rests on premisses that cannot be known with certainty. Even if the premisses are assumed to be true and the inference is a valid inductive inference, the conclusion may be false. The most we can say is that, with respect to given premisses, the conclusion has a certain degree of probability. Inductive logic tells us how to calculate the value of this probability.

We know that singular statements of fact, obtained by observation, are never absolutely certain because we may make errors in our observations; but, in respect to laws, there is still greater uncertainty. A law about the world states that, in any particular case, at any place and any time, if one thing is true, another thing is true. Clearly, this speaks about an infinity of possible instances. The actual instances may not be infinite, but there is an infinity of possible instances. A physiological law says that, if you stick a dagger into the heart of any human being, he will die. Since no exception to this law has ever been observed, it is accepted as universal. It is true, of course, that the number of instances so far observed of daggers being thrust into human hearts is finite. It is possible that some day humanity may cease to exist; in that case, the number of human beings, both past and future, is finite. But we do not know that humanity will cease to exist. Therefore, we must say that there is an infinity of possible instances, all of which are covered by the law. And, if there is an infinity of instances, no number of finite observations, however large, can make the "universal" law certain.

Of course, we may go on and make more and more observations, making them in as careful and scientific a manner as we can, until eventually we may say: "This law has been tested so many times that we can have complete confidence in its truth. It is a well-established, well-founded law." If we think about it, however, we see that even the best-founded laws of physics must rest on only a finite number of observations. It is always possible that tomorrow a counterinstance may be found. At no time is it possible to arrive at *complete* verification of a law. In fact, we should not speak of "verification" at all—if by the word we mean a definitive establishment of truth—but only of confirmation.

Interestingly enough, although there is no way in which a law can be verified (in the strict sense), there is a simple way it can be falsified. One need find only a single counterinstance. The knowledge of a counterinstance may, in itself, be uncertain. You may have made an error of observation or have been deceived in some way. But, if we assume that the counterinstance is a fact, then the negation of the law follows immediately. If a law says that every object that is *P* is also *Q* and we find an object that is *P* and not *Q*, the law is refuted. A million positive instances are insufficient to verify the law; one counterinstance is sufficient to falsify it. The situation is strongly asymmetric. It is easy to refute a law; it is exceedingly difficult to find strong confirmation.

How do we find confirmation of a law? If we have observed a great many positive instances and no negative instance, we say that the confirmation is strong. How strong it is and whether the strength can be expressed numerically is still a controversial question in the philosophy of science. We will return to this in a moment. Here we are concerned only with making clear that our first task in seeking confirmation of a law is to test instances to determine whether they are positive or negative. This is done by using our logical schema to make predictions. A law states that $(x) (Px \supset Qx)$; hence, for a given object a, $Pa \supset Qa$. We try to find as many objects as we can (here symbolized by $"a"$) that have the property P. We then observe whether they also fulfill the condition Q. If we find a negative instance, the matter is settled. Otherwise, each positive instance is additional evidence adding to the strength of our confirmation.

There are, of course, various methodological rules for efficient testing. For example, instances should be diversified as much as possible. If you are testing the law of thermal expansion, you should not limit your tests to solid substances. If you are testing the law that all metals are good conductors of electricity, you should not confine your tests to specimens of copper. You should test as many metals as possible under various conditions—hot, cold, and so on. We will not go into the many methodological rules for testing; we will only point out that in all cases the law is tested by making predictions and then seeing whether those predictions hold. In some cases, we find in nature the objects that we wish to test. In other cases, we have to produce them. In testing the law of thermal expansion, for example, we do not look for objects that are hot; we take certain objects and heat them. Producing conditions for testing has the great advantage that we can more easily follow the methodological rule of diversification; but whether we create the situations to be tested or find them ready-made in nature, the underlying schema is the same.

A moment ago I raised the question of whether the degree of confirmation of a law (or a singular statement that we are predicting by means of the law) can be expressed in quantitative form. Instead of saying that one law is "well founded" and that another law "rests on flimsy evidence", we might say that the first law has a .8 degree of confirmation, whereas the degree of confirmation for the second law is only .2. This question has long been debated. My own view is that such a procedure is legitimate and that what I have called "degree of confirmation" is identical with logical probability.

Such a statement does not mean much until we know what is meant by "logical probability". Why do I add the adjective "logical"? It is not customary practice; most books on probability do not make a distinction between various kinds of probability, one of which is called "logical". It is my belief, however, that there are two fundamentally different kinds of probability, and I distinguish between them by calling one "statistical probability", and the other "logical probability". It is unfortunate that the same word, "probability", has been used in two such widely differing senses. Failing to make the distinction is a source of enormous confusion in books on the philosophy of science as well as in the discourse of scientists themselves.

Instead of "logical probability", I sometimes use the term "inductive probability", because in my conception this is the kind of probability that is meant whenever we make an inductive inference. By "inductive inference" I mean, not only inference from facts to laws, but also any inference that is "nondemonstrative"; that is, an inference such that the conclusion does not follow with logical necessity when the truth of the premises is granted. Such inferences must be expressed in degrees of what I call "logical probability" or "inductive probability". To see clearly the distinction between this type of probability and statistical probability, it will be useful to glance briefly at the history of probability theory.

The first theory of probability, now usually called the "classical theory", was developed during the eighteenth century. Jacob Bernoulli (1654–1705) was the first to write a systematic treatise about it, and the Reverend Thomas Bayes made an important contribution. Toward the end of the century, the great mathematician and physicist Pierre Simon de Laplace wrote the first large treatise on the subject. It was a comprehensive mathematical elaboration of a theory of probability and may be regarded as the climax of the classical period.

The application of probability throughout the classical period was mostly to such games of chance as dice, cards, and roulette. Actually, the theory had its origin in the fact that some gamblers of the time had asked Pierre Fermat and other mathematicians to calculate for them the exact probabilities involved in certain games of chance. So the theory began with concrete problems, not with a general mathematical theory. The mathematicians found it strange that questions of this sort could be answered even though there was no field of mathematics available for providing such answers. As a consequence, they developed the theory of combinatorics, which could then be applied to problems of chance.

What did these men who developed the classical theory understand by "probability"? They proposed a definition that is still found in elementary books on probability: probability is the ratio of the number of favorable cases to the number of all possible cases. Let us see how this works in a simple example. Someone says: "I will throw this die. What is the chance that I will throw either an ace or a deuce?" The answer, according to the classical theory, is as follows. There are two "favorable" cases, that is, cases that fulfill the conditions specified in the question. Altogether, there are six possible ways the die can fall. The ratio of favorable to possible cases is therefore 2:6 or 1:3. We answer the question by saying that there is a probability of ⅓ that the die will show either a deuce or an ace.

All this seems quite clear, even obvious, but there is one important hitch to the theory. The classical authors said that, before one can apply their definition of probability, it must be ensured that all the cases involved are equally probable. Now we seem trapped in a vicious circle. We attempt to define what we mean by probability, and in so doing we use the concept of "equally probable". Actually, proponents of the classical theory did not put it in just those terms. They said that the cases must be "equipossible". This in turn was defined by a famous principle that they called "the principle of insufficient reason". Today it is usually called "the principle of indifference". If you do not know of any reason why one case should occur rather than another, then the cases are equipossible.

Such, in brief, was the way probability was defined in the classical period. A comprehensive mathematical theory has been built on the classical approach, but the only question that concerns us here is whether the foundation of this theory—the classical definition of probability—is adequate for science.

Slowly, during the nineteenth century, a few critical voices were raised against the classical definition. In the twentieth century, about 1920, both Richard von Mises and Hans Reichenbach made strong criticisms of the classical approach.[1] Mises said that "equipossibility" cannot be understood except in the sense of "equiprobability". If this is what it means, however, we are indeed caught in a vicious circle. The classical tradition, Mises asserted, is circular and therefore unusable.

Mises had still another objection. He granted that, in certain simple cases, we can rely on common sense to tell us that certain events are equipossible. We can say that heads and tails are equipossible outcomes when a coin is flipped because we know of no reason why one should turn up rather than the other. Similarly with roulette; there is no reason why the ball should fall into one compartment rather than another. If playing cards are of the same size and shape, with identical backs, and are well shuffled, then one card is as likely to be dealt to a player as any other. Again, the conditions of equipossibility are fulfilled. But, Mises went on, none of the classical authors pointed out how this definition of probability could be applied to many other situations. Consider mortality tables. Insurance companies have to know the probability that a forty-year-old man, in the United States, with no serious diseases will live to the same date in the following year. They must be able to calculate probabilities of this sort because they are the basis on which the company determines its rates.

What, Mises asked, are the equipossible cases for a man? Mr. Smith applies for life insurance. The company sends him to a doctor. The doctor reports that Smith has no serious diseases and that his birth certificate shows him to be forty years old. The company looks at its mortality tables; then, on the basis of the man's probable life expectancy, it offers him insurance at a certain rate. Mr. Smith may die before

---

[1] On the views of Mises and Reichenbach, see Richard von Mises, *Probability, Statistics, and Truth* (New York: Macmillan, 1939), and Hans Reichenbach, *The Theory of Probability* (Berkeley, Calif.: University of California Press, 1949).

he reaches forty-one, or he may live to be a hundred. The probability of surviving one more year goes down and down as he gets older. Suppose he dies at forty-five. This is bad for the insurance company because he paid only a few premiums, and now they have to pay $20,000 to his beneficiary. Where are the equipossible cases? Mr. Smith may die at the age of forty, of forty-one, of forty-two, and so on. These are the possible cases. But they are not equipossible; that he will die at the age of 120 is extremely improbable.

A similar situation prevails, Mises pointed out, in applying probability to the social sciences, to weather prediction, and even to physics. These situations are not like games of chance, in which the possible outcomes can be classified neatly into n mutually exclusive, completely exhaustive cases that fulfill the conditions of equipossibility. A small body of radioactive substance will, in the next second, either emit an alpha particle or it will not. The probability that it will emit the particle is, say, .0374. Where are the equipossible cases? There are none. We have only two cases: either it will emit the alpha particle in the next second or it will not emit it. This was Mises' chief criticism of the classical theory.

On the constructive side, both Mises and Reichenbach had this to say. What we really mean by probability has nothing to do with counting cases. It is a measurement of "relative frequency". By "absolute frequency", we mean the total number of objects or occurrences; for example, the number of people in Los Angeles who died last year of tuberculosis. By "relative frequency", we mean the ratio of this number to that of a larger class being investigated, say, the total number of inhabitants of Los Angeles.

We can speak of the probability that a certain face of a die will be thrown, Mises said, not only in the case of a fair die, where it is ⅙, but also in cases of all types of loaded dice. Suppose someone asserts that the die he has is loaded and that the probability it will show an ace is not ⅙, but less than ⅙. Someone else says: "I agree with you that the die is loaded, but not in the way you believe. I think that the probability of an ace is greater than ⅙." Mises pointed out that, in order to learn what the two men mean by their divergent assertions, we must look at the way they try to settle their argument. They will, of course, make an empirical test. They will toss the die a number of times, keeping a record of the number of throws and the number of aces.

How many times will they toss the die? Suppose they make 100 throws and find that the ace comes up 15 times. This is slightly less than ⅙ of 100. Will this not prove that the first man is right? "No", the other man might say. "I still think the probability is greater than ⅙. One hundred throws is not sufficient for an adequate test." Perhaps the men continue tossing the die until they have made 6,000 throws. If the ace has turned up fewer than 1,000 times, the second man may decide to give up. "You are right", he says. "It is less than ⅙."

Why do the men stop at 6,000? It may be that they are tired of making throws. Perhaps they made a bet of a dollar about which way the die was loaded, and for a mere dollar they do not want to spend three more days on additional throws. But the decision to stop at 6,000 is purely arbitrary. If, after 6,000 throws, the number of aces is very close to 1,000, they might regard the question as still undecided. A small deviation could be due to chance, rather than to a physical bias in the die itself. In a longer run, the bias might cause a deviation in the opposite direction. To make a more decisive test, the men might decide to go on to 60,000 throws. Clearly, there is no finite number of throws, however large, at which they could stop the test and say with positive assurance that the probability of an ace is ⅙ or less than ⅙ or more.

Since no finite number of tests is sufficient for determining a probability with certainty, how can that probability be defined in terms of frequency? Mises and Reichenbach proposed that it be defined, not as relative frequency in a finite series of instances, but as the *limit* of the relative frequency in an endless series. (It was this definition that distinguished the views of Mises and Reichenbach from those of R.A. Fisher, in England, and other statisticians who had also criticized the classical theory. They introduced the frequency concept of probability, not by definition, but as a primitive term in an axiom system.) Of course, Mises and Reichenbach were well aware—although they have often been criticized as though they had not been—that no observer can ever have the complete infinite series of observations available. But I think that their critics were wrong when they said that the new definition of probability has no application. Both Reichenbach and Mises have shown that many theorems can be developed on the basis of their definition, and, with the help of these theorems, we can say something significant. We cannot say with certainty what the value of a probability is, but, if the series is long enough, we can say what the probability *probably* is. In the die example, we might say that the probability that the probability of throwing an ace is greater than ⅙ is very small. Perhaps the value of this probability of a probability can even be calculated. The facts that the limit concept is used in the definition and that reference is made to an infinite series certainly do cause complications and difficulties, both logical and practical. They do not, however, make the definition meaningless, as some critics have asserted.

Reichenbach and Mises agreed in the view that this concept of probability, based on the limit of a relative frequency in an infinite series, is the only concept of probability acceptable in science. The classical definition, derived from the principle of indifference, had been found inadequate. No new definition other than that of Mises and Reichenbach had been found that was superior to the old. But now the troublesome question of single instances arose once more. The new definition worked very well for statistical phenomena, but how could it be applied to a single case? A meteorologist announces that the probability of rain tomorrow is ⅔. "Tomorrow" refers to one particular day and no other. Like the death of the man applying for life insurance, it is a single, unrepeated event; yet we want to attribute to it a probability. How can this be done on the basis of a frequency definition?

Mises thought that it could not be done; therefore, probability statements for single cases should be excluded. Reichenbach, however, was aware that, in both science and everyday life, we constantly make probability statements about single events. It would be useful, he thought, to find a plausible interpretation for such statements. In weather prediction, it is easy to give such an interpretation. The meteorologist has available a large number of reports of past observations of the weather, as well as data concerning the weather for today. He finds that today's weather belongs to a certain class, and that in the past, when weather of this class occurred, the relative frequency with which rain fell on the following day was ⅔. Then, according to Reichenbach, the meteorologist makes a "posit"; that is, he assumes that the observed frequency of ⅔, based on a finite but rather long series of observations, is also the limit of the infinite series. In other words, he estimates the limit to be in the neighborhood of ⅔. He then makes the statement: "The probability of rain tomorrow is ⅔."

The meteorologist's statement, Reichenbach maintained, should be regarded as an elliptical one. If he expanded it to its full meaning, he would say: "According to our past observations, states of weather such as that we have observed today were followed, with a frequency of ⅔, by rain on the following day." The abbreviated statement seems to apply probability to a single case, but that is only a manner of speaking. The statement really refers to relative frequency in a long series. The same would be true of the statement: "On the next throw of the die, the probability of an ace is $\frac{1}{6}$." The "next throw" is, like "the weather tomorrow", a single, unique event. When we attribute probability to it, we are really speaking elliptically about relative frequency in a long series of throws.

In this way, Reichenbach found an interpretation for statements that attributed probability to single events. He even tried to find an interpretation for statements attributing probability to general hypotheses in science. We will not enter into that here because it is more complicated and because (in contrast to his interpretation of singular probability predictions) it has not found general acceptance.

The next important development in the history of probability theory was the rise of the *logical* conception. It was proposed after 1920 by John Maynard Keynes, the famous British economist, and has since been elaborated by many writers. Today there is a spirited controversy between proponents of this logical conception and those who favor the frequency interpretation. The next chapter will discuss this controversy and the manner in which I think it should be resolved.

CHAPTER **3**

Induction and
Logical Probability

TO JOHN MAYNARD KEYNES, proba-
bility was a logical relation between two propositions. He did not try
to define this relation. He even went so far as to say that no definition
could be formulated. Only by intuition, he insisted, can we understand
what probability means. His book, *A Treatise on Probability*,[1] gave a
few axioms and definitions, expressed in symbolic logic, but they are
not very sound from a modern point of view. Some of Keynes's axioms
were actually definitions. Some of his definitions were really axioms.
But his book is interesting from a philosophical standpoint, especially
those chapters in which he discusses the history of probability theory
and what can be learned today from earlier points of view. His central
contention was that, when we make a probability statement, we are not
making a statement about the world, but only about a logical relation
between two other statements. We are saying only that one statement
has a logical probability of so-and-so much with respect to another
statement.

I use the phrase "so-and-so much". Actually, Keynes was more

[1] John Maynard Keynes, *Treatise on Probability* (London: Macmillan, 1921).

cautious. He doubted that probability in general could be made a quantitative concept, that is, a concept with numerical values. He agreed, of course, that this could be done in special cases, such as the throw of a die, in which the old principle of indifference applied. The die is symmetrical, all its faces are alike, we have no reason to suspect it is loaded, and so on. The same is true of other games of chance, in which conditions are carefully arranged to produce physical symmetry, or, at least, symmetry with respect to our knowledge and ignorance. Roulette wheels are made so that their various sectors are equal. The wheel is carefully balanced to eliminate any bias that might cause the ball to stop at one number rather than another. If someone flips a coin, we have no reason to suppose that heads will show rather than tails.

In restricted situations of this sort, Keynes said, we can legitimately apply something like the classical definition of probability. He agreed with other critics of the principle of indifference that it had been used in the classical period in much too wide a sense and that it had been wrongly applied to many situations, such as the prediction that tomorrow the sun will rise. It is true, he said, that in games of chance and other simple situations, the principle of indifference is applicable, and numerical values can be given to probability. In most situations, however, we have no way of defining equipossible cases and, therefore, no justification for applying the principle. In such cases, Keynes said, we should not use numerical values. His attitude was cautious and skeptical. He did not want to go too far, to tread on what he regarded as thin ice, so he restricted the quantitative part of his theory. In many situations in which we do not hesitate to make bets, to attribute numerical values to probability predictions, Keynes cautioned against the practice.

The second important figure in the rise of the modern logical approach to probability is Harold Jeffreys, an English geophysicist. His *Theory of Probability,* first published in 1939 by Oxford Press, defends a conception closely related to that of Keynes. When Keynes published his book (it came out in 1921, so he probably wrote it in 1920), the very first publications on probability by Mises and Reichenbach had just appeared. Keynes apparently did not know about them. He criticized the frequency approach, but he did not discuss it in detail. By the time Jeffreys wrote his book, the frequency interpretation had been fully developed, so his book was much more explicit in dealing with it.

Jeffreys said flatly that the frequency theory is entirely wrong. He affirmed Keynes's view that probability refers not to frequency but to a logical relation. He was much more daring than the cautious Keynes. He believed that numerical values *could* be assigned to probability in a large number of situations, especially in all those situations in which mathematical statistics is applied. He wanted to deal with the same problems that interested R. A. Fisher and other statisticians, but he wanted to deal with them on the basis of a different concept of probability. Because he made use of an indifference principle, I believe that some of his results are open to the same objections that were raised against the classical theory. It is difficult, however, to find specific statements in his book to criticize. His axioms, taken one after the other, are acceptable. Only when he tries to derive theorems from one certain axiom does he, in my opinion, go astray.

The axiom in question is stated by Jeffreys as follows: "We assign the larger number on given data to the more probable proposition (and therefore equal numbers to equally probable propositions)." The part included in the parenthesis obviously says only that, if *p* and *q* are equally probable on the basis of evidence *r*, then equal numbers are to be assigned to *p* and *q* as their probability values with respect to evidence *r*. The statement tells us nothing about the conditions under which we are to regard *p* and *q* as equally probable with respect to *r*. Nowhere else in his book does Jeffreys state those conditions. Later in his book, however, he interprets this axiom in a most surprising way in order to establish theorems about scientific laws. "If there is no reason to believe one hypothesis rather than another", he writes, "the probabilities are equal." In other words, if we have insufficient evidence for deciding whether a given theory is true or false, we must conclude that the theory has a probability of ½.

Is this a legitimate use of the principle of indifference? In my view, it is a use that was rightly condemned by critics of the classical theory. If the principle of indifference is to be used at all, there must be some sort of symmetry in the situation, such as the equality of the faces of a die or the sectors of a roulette wheel, that enables us to say that certain cases are equally probable. In the absence of such symmetries in the logical or physical features of a situation, it is unwarranted to assume equal probabilities merely because we know nothing about the relative merits of rival hypotheses.

A simple illustration will make this clear. According to Jeffreys' interpretation of his axiom, we could assume a probability of ½ that there are living organisms on Mars because we have neither sufficient reason to believe this hypothesis nor sufficient reason to believe its negation. In the same way, we could reason that the probability is ½ that there are animals on Mars and ½ that there are human beings there. Each assertion, considered by itself, is an assertion about which we have no sufficient evidence one way or the other. But these assertions are related to each other in such a way that they cannot have the same probability values. The second assertion is stronger than the first because it implies the first, whereas the first does not imply the second. Therefore, the second assertion has less probability than the first; the same relation holds between the third and the second. We must be extremely careful, therefore, in applying even a modified principle of indifference, or we are likely to run into such inconsistencies.

Jeffreys' book has been harshly criticized by mathematical statisticians. I agree with their criticism only with respect to the few places where Jeffreys develops theorems that cannot be derived from his axioms. On the other hand, I would say that both Keynes and Jeffreys were pioneers who worked in the right direction.[2] My own work on probability is in the same direction. I share their view that logical probability is a logical relation. If you make a statement affirming that, for a given hypothesis, the logical probability with respect to given evidence is .7, then the total statement is an analytic one. This means that the statement follows from the definition of logical probability (or from the axioms of a logical system) without reference to anything outside the logical system, that is, without reference to the structure of the actual world.

In my conception, logical probability is a logical relation somewhat similar to logical implication; indeed, I think probability may be regarded as a partial implication. If the evidence is so strong that the hypothesis follows logically from it—is logically implied by it—we have one extreme case in which the probability is 1. (Probability 1 also occurs in other cases, but this is one special case where is occurs.) Similarly, if the negation of a hypothesis is logically implied by the evidence, the logical probability of the hypothesis is 0. In between, there is a continuum of cases about which deductive logic tells us nothing beyond the negative assertion that neither the hypothesis nor its negation can be deduced from the evidence. On this continuum inductive logic must take

---

[2] A technical evaluation of the work of Keynes and Jeffreys, and others who defended logical probability, will be found in section 62 of my *Logical Foundations of Probability* (Chicago: University of Chicago Press, 1950). Six nontechnical sections of this book were reprinted as a small monograph, *The Nature and Application of Inductive Logic* (Chicago: University of Chicago Press, 1951).

over. But inductive logic is like deductive logic in being concerned solely with the statements involved, not with the facts of nature. By a logical analysis of a stated hypothesis *h* and stated evidence *e*, we conclude that *h* is not logically implied but is, so to speak, partially implied by *e* to the degree of so-and-so much.

At this point, we are justified, in my view, in assigning numerical value to the probability. If possible, we should like to construct a system of inductive logic of such a kind that for any pair of sentences, one asserting evidence *e* and the other stating a hypothesis *h*, we can assign a number giving the logical probability of *h* with respect to *e*. (We do not consider the trivial case in which the sentence *e* is contradictory; in such instances, no probability value can be assigned to *h*.) I have succeeded in developing possible definitions of such probabilities for very simple languages containing only one-place predicates, and work is now in progress for extending the theory to more comprehensive languages. Of course, if the whole of inductive logic, which I am trying to construct on this basis, is to be of any real value to science, it should finally be applicable to a quantitative language such as we have in physics, in which there are not only one- or two-place predicates, but also numerical magnitudes such as mass, temperature, and so on. I believe that this is possible and that the basic principles involved are the same as the principles that have guided the work so far in the construction of an inductive logic for the simple language of one-place predicates.

When I say I think it is possible to apply an inductive logic to the language of science, I do not mean that it is possible to formulate a set of rules, fixed once and for all, that will lead automatically, in any field, from facts to theories. It seems doubtful, for example, that rules can be formulated to enable a scientist to survey a hundred thousand sentences giving various observational reports and then find, by a mechanical application of those rules, a general theory (system of laws) that would explain the observed phenomena. This is usually not possible, because theories, especially the more abstract ones dealing with such nonobservables as particles and fields, use a conceptual framework that goes far beyond the framework used for the description of observation material. One cannot simply follow a mechanical procedure based on fixed rules to devise a new system of theoretical concepts, and with its help a theory. Creative ingenuity is required. This point is sometimes expressed by saying that there cannot be an inductive machine—a computer into which we can put all the relevant observational sentences and get, as an output, a neat system of laws that will explain the observed phenomena.

I agree that there cannot be an inductive machine if the purpose of the machine is to invent new theories. I believe, however, that there can be an inductive machine with a much more modest aim. Given certain observations *e* and a hypothesis *h* (in the form, say, of a prediction or even of a set of laws), then I believe it is in many cases possible to determine, by mechanical procedures, the logical probability, or degree of confirmation, of *h* on the basis of *e*. For this concept of probability, I also use the term "inductive probability", because I am convinced that this is the basic concept involved in all inductive reasoning and that the chief task of inductive reasoning is the evaluation of this probability.

When we survey the present situation in probability theory, we find a controversy between advocates of the frequency theory and those who, like Keynes, Jeffreys, and myself, speak in terms of a logical probability. There is, however, one important difference between my position and that of Keynes and Jeffreys. They reject the frequency concept of probability. I do not. I think the frequency concept, also called statistical probability, is a good scientific concept, whether introduced by an explicit definition, as in the systems of Mises and Reichenbach, or introduced by an axiom system and rules of practical application (without explicit definition), as in contemporary mathematical statistics. In both cases, I regard this concept as important for science. In my opinion, the logical concept of probability is a second concept, of an entirely different nature, though equally important.

Statements giving values of statistical probability are not purely logical; they are factual statements in the language of science. When a medical man says that the probability is "very good" (or perhaps he uses a numerical value and says .7) that a patient will react positively to a certain injection, he is making a statement in medical science. When a physicist says that the probability of a certain radioactive phenomenon is so-and-so much, he is making a statement in physics. Statistical probability is a scientific, empirical concept. Statements about statistical probability are "synthetic" statements, statements that cannot be decided by logic but which rest on empirical investigations. On this point I agree fully with Mises, Reichenbach, and the statisticians. When we say, "With this particular die the statistical probability of throwing an ace is .157", we are stating a scientific hypothesis that can be tested only by a series of observations. It is an empirical statement because only an empirical investigation can confirm it.

As science develops, probability statements of this sort seem to become increasingly important, not only in the social sciences, but in modern physics as well. Statistical probability is involved not only in areas where it is necessary because of ignorance (as in the social sciences or when a physicist is calculating the path of a molecule in a liquid), but also as an essential factor in the basic principles of quantum theory. It is of the utmost importance for science to have a theory of statistical probability. Such theories have been developed by statisticians and, in a different way, by Mises and Reichenbach.

On the other hand, we also need the concept of logical probability. It is especially useful in metascientific statements, that is, statements about science. We say to a scientist: "You tell me that I can rely on this law in making a certain prediction. How well established is the law? How trustworthy is the prediction?" The scientist today may or may not be willing to answer a metascientific question of this kind in quantitative terms. But I believe that, once inductive logic is sufficiently developed, he could reply: "This hypothesis is confirmed to degree .8 on the basis of the available evidence." A scientist who answers in this way is making a statement about a logical relation between the evidence and the hypothesis in question. The sort of probability he has in mind is logical probability, which I also call "degree of confirmation". His statement that the value of this probability is .8 is, in this context, not a synthetic (empirical) statement, but an analytic one. It is analytic because no empirical investigation is demanded. It expresses a logical relation between a sentence that states the evidence and a sentence that states the hypothesis.

Note that, in making an analytic statement of probability, it is always necessary to specify the evidence explicitly. The scientist must not say: "The hypothesis has a probability of .8." He must add, "with respect to such and such evidence." If this is not added, his statement might be taken as a statement of statistical probability. If he intends it to be a statement of logical probability, it is an elliptical statement in which an important component has been left out. In quantum theory, for instance, it is often difficult to know whether a physicist means statistical probability or logical probability. Physicists usually do not draw this distinction. They talk as though there were only one concept of probability with which they work. "We mean that kind of probability that fulfills the ordinary axioms of probability theory", they may say. But the ordinary axioms of probability theory are fulfilled by both concepts, so this remark does not clear up the question of exactly what type of probability they mean.

A similar ambiguity is found in the statements of Laplace and others who developed the classical conception of probability. They were not aware, as we are today, of the difference between logical probability and frequency probability. For that reason it is not always possible to determine which concept they meant. I am convinced, however, that most of the time—not always, of course—they meant the logical concept. Mises and other frequentists were not correct, in my opinion, in certain criticisms they made of the classical school. Mises believed that there was no other scientific concept of probability but the frequency concept, so he assumed that, if the classical writers meant anything at all by "probability", they must have meant statistical probability. Of course, they were not able to say clearly and explicitly that they meant relative frequency in the long run, but this, according to Mises, is what they implicitly meant. I do not agree. I believe that, when the classical writers made certain statements about a priori probability, they were speaking of logical probability, which is analytic and therefore *can* be known a priori. I do not regard these statements as violations of the principle of empiricism, as Mises and Reichenbach do.

Let me add a word of caution. After I had expressed this view in my book on probability, a number of colleagues—some of them my friends—pointed to certain quotations from classical authors and said that logical probability could not have been what those authors had in mind. With this I agree. In some of their statements the classical writers could not have meant logical probability; presumably, they meant frequency probability. Nevertheless, I am convinced that their basic concept was logical probability. I think this is even implied by the title of the first systematic book in the field, Jacob Bernoulli's *Ars conjectandi,* the art of conjecture. Mises' theory of probability is not an art of conjecture. It is a mathematically formulated axiomatic theory of mass phenomena. There is nothing conjectural about it. What Bernoulli meant was quite different. We have seen certain events, he said, such as the way a die has fallen, and we want to make a conjecture about how it will fall if we throw it again. We want to know how to make rational bets. Probability, for the classical writers, was the degree of certainty or confidence that our beliefs can have about future events. This is logical probability, not probability in the statistical sense.[3]

[3] My general view, that both statistical and logical probability are legitimate, good scientific concepts that play different roles, is expressed in Chapter II of *Logical Foundations of Probability,* cited in the previous footnote, and in my 1945 paper, "The Two Concepts of Probability," reprinted in Herbert Feigl and Wilfrid Sellars, eds., *Readings in Philosophical Analysis* (New York: Appleton-Century-

I will not go into greater detail here about my view of probability, because many technicalities are involved. But I will discuss the one inference in which the two concepts of probability may come together. This occurs when either the hypothesis or one of the premises for the inductive inference contains a concept of statistical probability. We can see this easily by modifying the basic schema used in our discussion of universal laws. Instead of a universal law (1), we take as the first premiss a statistical law (1'), which says that the relative frequency (*rf*) of *Q* with respect to *P* is (say) .8. The second premiss (2) states, as before, that a certain individual *a* has the property *P*. The third statement (3) asserts that *a* has the property *Q*. This third statement, *Qa*, is the hypothesis we wish to consider on the basis of the two premises.

In symbolic form:

$$(1') \quad rf(Q,P) = .8$$
$$(2) \quad Pa$$
$$(3) \quad Qa$$

What can we say about the logical relation of (3) to (1') and (2)? In the previous case—the schema for a universal law—we could make the following logical statement:

(4) Statement (3) is logically implied by (1) and (2).

We cannot make such a statement about the schema given above because the new premiss (1') is weaker than the former premiss (1); it states a relative frequency rather than a universal law. We *can*, however, make the following statement, which also asserts a logical relation, but in terms of logical probability or degree of confirmation, rather than in terms of implication:

(4') Statement (3), on the basis of (1') and (2), has a probability of .8.

Note that this statement, like statement (4), is not a logical inference from (1') and (2). Both (4) and (4') are statements in what is called a metalanguage; they are logical statements *about* three assertions: (1) [or (1'), respectively], (2), and (3).

It is important to understand precisely what is meant by such a statement as "The statistical probability of *Q* with respect to *P* is .8." When scientists make such statements, speaking of probability in the

Crofts, 1949), pp. 330–348, and Herbert Feigl and May Brodbeck, eds., *Readings in the Philosophy of Science* (New York: Appleton-Century-Crofts, 1953), pp. 438–455. For a more popularly written defense of the same viewpoint, see my article "What is Probability?," *Scientific American*, 189 (September 1953).

frequency sense, it is not always clear exactly what frequency they mean. Is it the frequency of Q in an observed sample? Is it the frequency of Q in the total population under consideration? Is it an *estimate* of the frequency in the total population? If the number of observed instances in the sample is very large, then the frequency of Q in the sample may not differ in any significant degree from the frequency of Q in the population or from an estimate of this frequency. Nevertheless, it is important to keep in mind the theoretical distinctions involved here.

Suppose that we wish to know what percentage of a hundred thousand men living in a certain city shave with electric razors. We decide to question one thousand of these men. To avoid a biased sample, we must select the thousand men in ways developed by workers in the field of modern polling techniques. Assume that we obtain an unbiased sample and that eight hundred men in the sample report that they use an electric razor. The observed relative frequency of this property is, therefore, .8. Since one thousand is a fairly large sample, we might conclude that the statistical probability of this property, in the total population, is .8. Strictly speaking, this is not a warranted conclusion. Only the value of the frequency in the sample is known. The value of the frequency in the population is not known. The best we can do is make an *estimate* of the frequency in the population. This estimate must not be confused with the value of the frequency in the sample. In general, such estimates should deviate in a certain direction from the observed relative frequency in a sample.[4]

Assume that (1') is known: the statistical probability of Q, with respect to P, is .8. (How we know this is a question that need not be considered. We may have tested the entire population of a hundred thousand by interviewing every man in the city.) The statement of this probability is, of course, an empirical statement. Suppose, also, that the second premiss is known: (2) *Pa*. We can now make statement (4'), which says that the logical probability of (3) *Qa*, with respect to premisses (1') and (2), is .8. If, however, the first premiss is not a statement of statistical probability, but the statement of an observed relative frequency in a sample, then we must take into consideration the size of the sample. We can still calculate the logical probability, or degree of confirmation, expressed in statement (4), but it will not be ex-

---

[4] This question is not discussed in my *Logical Foundations of Probability;* but in a small monograph, *The Continuum of Inductive Methods* (University of Chicago Press, 1952), I have developed a number of techniques for estimating relative frequency on the basis of observed samples.

actly .8. It will deviate in ways I have discussed in the monograph mentioned in the previous footnote.

When an inductive inference is made in this way, from a sample to the population, from one sample to an unknown future sample, or from one sample to an unknown future instance, I speak of it as "indirect probability inference" or "indirect inductive inference", as distinct from the inductive inference that goes from the population to a sample or an instance. As I have said earlier, *if* knowledge of the actual statistical probability in the population is given in (1′), it is correct to assert in (4) the same numerical value for the degree of confirmation. Such an inference is not deductive; it occupies a somewhat intermediate position between the other kinds of inductive and deductive inferences. Some writers have even called it a "deductive probability inference", but I prefer to speak of it as inductive rather than deductive. Whenever the statistical probability for a population is given and we wish to determine the probability for a sample, the values given by my inductive logic are the same as those a statistician would give. If, however, we make an indirect inference from a sample to the population or from a sample to a future single instance or a future finite sample (these two latter cases I call "predictive inferences"), then I believe that the methods used in statistics are not quite adequate. In my monograph on *The Continuum of Inductive Methods,* I give in detail the reasons for my scepticism.

The main points that I wish to stress here are these: Both types of probability—statistical and logical—may occur together in the same chain of reasoning. Statistical probability is part of the object language of science. To statements about statistical probability we can apply logical probability, which is part of the metalanguage of science. It is my conviction that this point of view gives a much clearer picture of statistical inference than is commonly found in books on statistics and that it provides an essential groundwork for the construction of an adequate inductive logic of science.

CHAPTER 4

The Experimental Method

ONE OF THE GREAT distinguishing features of modern science, as compared to the science of earlier periods, is its emphasis on what is called the "experimental method". As we have seen, all empirical knowledge rests finally on observations, but these observations can be obtained in two essentially different ways. In the non-experimental way, we play a passive role. We simply look at the stars or at some flowers, note similarities and differences, and try to discover regularities that can be expressed as laws. In the experimental way, we take an active role. Instead of being onlookers, we *do* something that will produce better observational results than those we find by merely looking at nature. Instead of waiting until nature provides situations for us to observe, we try to create such situations. In brief, we make experiments.

The experimental method has been enormously fruitful. The great progress physics has made in the last two hundred years, especially in the last few decades, would have been impossible without the experimental method. If this is so, one might ask, why is the experimental method not used in all fields of science? In some fields it is not as easy to use as it is in physics. In astronomy, for example, we cannot give a planet a push in some other direction to see what would happen to it. Astronomical objects are out of reach; we can only observe and describe them. Sometimes astronomers can create conditions in the laboratory similar to those, say, on the surface of the sun or moon and then observe what happens in the laboratory under those conditions. But this is not really an astronomical experiment. It is a physical experiment that has some relevance for astronomical knowledge.

Entirely different reasons prevent social scientists from making experiments with large groups of people. Social scientists do make experiments with groups, but usually they are small groups. If we want to learn how people react when they are unable to obtain water, we can take two or three people, give them a diet without liquid, and observe their reactions. But this does not tell us much about how a large community would react if its water supply were cut off. It would be an interesting experiment to stop the water supply to New York, for instance. Would people get frantic or apathetic? Would they try to organize a revolution against the city government? Of course, no social scientist would suggest making such an experiment because he knows that the community would not permit it. People will not allow social scientists to play with their essential needs.

Even when no real cruelty to a community is involved, there are often strong social pressures against group experiments. For example, there is a tribe in Mexico that performs a certain ritual dance whenever there is an eclipse of the sun. Members of the tribe are convinced that only in this way can they placate the god who is causing the eclipse. Finally, the light of the sun returns. Suppose a group of anthropologists tries to convince these people that their ritual dance had nothing to do with the sun's return. The anthropologists propose that the tribe experiment by not doing the dance the next time the sun's light goes away and seeing what happens. The tribesmen would respond with indignation. To them it would mean running the risk of living the rest of their days in darkness. They believe so strongly in their theory that they do not want to put it to a test. So, you see, there are obstacles to experiments in the social sciences even when the scientists are convinced that no social harm will result if the experiments are performed. The social scientist is, in general, restricted to what he can learn from history and from experiments with individuals and small groups. In a dictatorship, however, large group experiments are often made, not just to test a theory, but rather because the government believes that a new procedure may work better than an old one. The government experiments on a grand scale in agriculture, economics, and so on. In a democracy, it is not possible to make such audacious experiments because, if they did not turn out well, the government would have to face an angry public at the next election.

The experimental method is especially fruitful in fields in which there are quantitative concepts that can be accurately measured. How does the scientist plan an experiment? It is hard to describe the general nature of experiments, because there are so many different kinds, but a few general features can be pointed out.

First of all, we try to determine the relevant factors involved in the phenomenon we wish to investigate. Some factors—but not too many—must be left aside as irrelevant. In an experiment in mechanics, for example, involving wheels, levers, and so on, we may decide to disregard friction. We know that friction is involved, but we think its influence is too small to justify complicating the experiment by considering it. Similarly, in an experiment with slow-moving bodies, we may choose to neglect air resistance. If we are working with very high velocities, such as a missile moving at a supersonic speed, we can no longer neglect air resistance. In short, the scientist leaves out only those factors whose influence on his experiment will, he thinks, be insignificant. Sometimes, in order to keep an experiment from being too complicated, he may even have to neglect factors he thinks may have important effects.

After having decided on the relevant factors, we devise an experiment in which some of those factors are kept constant while others are permitted to vary. Suppose we are dealing with a gas in a vessel, and we wish to keep the temperature of the gas as constant as we can. We immerse the vessel in a water-bath of much larger volume. (The specific heat of the gas is so small in relation to the specific heat of the water that, even if the temperature of the gas is varied momentarily, as by compression or expansion, it will quickly go back to its old temperature.) Or we may wish to keep a certain electrical current at a constant rate of flow. Perhaps this is done by having an ammeter so that, if we observe an increase or decrease in the current, we can alter the resistance and keep the current constant. In such ways as these we are able to keep certain magnitudes constant while we observe what happens when other magnitudes are varied.

Our final aim is to find laws that connect *all* the relevant magnitudes; but, if a great many factors are involved, this may be a complicated task. At the beginning, therefore, we restrict our aim to lower-level laws that connect *some* of the factors. The simplest first step, if there are k magnitudes involved, is to arrange the experiment so that k-2 magnitudes are kept constant. This leaves two magnitudes, M_1 and M_2, that we are free to vary. We change one of them and observe how the other behaves. Maybe M_2 goes down whenever M_1 is increased. Or perhaps, as M_1 is increased, M_2 goes first up and then down. The value of M_2 is a function of the value of M_1. We can plot this function as a curve on a sheet of graph paper and perhaps determine the equation that expresses the function. We will then have a restricted law: If magnitudes M_3, M_4, M_5 . . . are kept constant, and M_1 is increased, M_2 varies in a way expressed by a certain equation. But this is only the beginning. We continue our experiment, controlling other sets of k-2 factors, so that we can see how other pairs of magnitudes are functionally related. Later, we experiment in the same way with triples, keeping everything constant except three magnitudes. In some cases, we may be able to guess, from our laws relating to pairs, some or all of the laws concerning the triples. Then we aim for still more general laws involving four magnitudes, and finally for the most general, sometimes quite complicated, laws that cover all the relevant factors.

As a simple example, consider the following experiment with a gas. We have made the rough observation that the temperature, volume, and pressure of a gas often vary simultaneously. We wish to know exactly how these three magnitudes are related to one another. A fourth relevant factor is what gas we are using. We may experiment with other gases later, but at first we decide to keep this factor constant by using only pure hydrogen. We put the hydrogen in a cylindrical vessel (see Figure 4–1) with a movable piston on which a weight can be placed. We can easily measure the volume of the gas, and we can vary the pressure by changing the weight on the piston. The temperature is regulated and measured by other means.

Before we proceed with experiments to determine how the three factors—temperature, volume, and pressure—are related, we need to make some preliminary experiments to make sure that there are no other relevant factors. Some factors we might suspect of being relevant turn out not to be. For example, is the shape of the vessel containing the gas relevant? We know that in some experiments (for example, the distribution of an electrical charge and its surface potential) the shape of the object involved is important. Here it is not difficult to determine that the shape of the vessel is irrelevant; only the volume is important. We can draw on our knowledge of nature to rule out many other factors. An astrologer may come into the laboratory and ask: "Have you checked on where the planets are today? Their positions may have some influence on your experiment." We consider this an irrelevant factor because we believe the planets are too far away to have an influence.

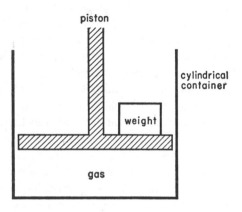

Figure 4–1.

Our assumption of the irrelevance of the planets is correct, but it would be a mistake to think that we can automatically exclude various factors simply because we believe they have no influence. There is no way to be really sure until experimental tests have been made. Imagine that you live before the invention of radio. Someone places a box on your table and tells you that if someone sings at a certain spot, one thousand miles away, you will hear the apparatus in this box sing exactly the same song, in the same pitch and rhythm. Would you believe it? You would probably reply: "Impossible! There are no electric wires attached to this box. I know from my experience that nothing happening one thousand miles away could have any effect on what is happening in this room."

That is exactly the same reasoning by which we decided that the positions of the planets could not affect our experiments with hydrogen! It is obvious that we must be very cautious. Sometimes there are influences we cannot know about until they are discovered. For this reason, the very first step in our experiment—determining the relevant factors—

is sometimes a difficult one. Moreover, it is a step that is often not explicitly mentioned in the reports of investigations. A scientist describes only the apparatus he used, the experiment he performed, what he discovered about the relations between certain magnitudes. He does not add, "and in addition I found out that such and such factors have no influence on the results". In most cases, when enough is known about the field in which the investigation is made, the scientist will take for granted that other factors are irrelevant. He may be quite right. But in new fields, one must be extremely cautious. Of course, nobody would think that a laboratory experiment could be influenced by whether we look at the apparatus from a distance of ten inches or ten feet or whether we are in a kind or angry disposition when we look at it. These factors are probably irrelevant, but absolutely sure we cannot be. If anyone suspects that these are relevant factors, an experiment must be made to rule them out.

Practical considerations prevent us, of course, from testing every factor that might be relevant. Thousands of remote possibilities can be tested, and there simply is not time to examine all of them. We must proceed according to common sense and correct our assumptions only if something unexpected happens that forces us to consider relevant a factor we had previously neglected. Will the color of leaves on trees outside a laboratory influence the wave length of light used in an experiment? Will a piece of apparatus function differently depending on whether its legal owner is in New York or Chicago or on how he feels about the experiment? We obviously do not have time to test such factors. We assume that the mental attitude of the equipment's owner has no physical influence on the experiment, but members of certain tribes may differ. They may believe that the gods will assist the experiment only if the owner of the apparatus wants the experiment made and not if the pretended owner wishes it. Cultural beliefs thus sometimes influence what is considered relevant. In most cases, a scientist thinks about the problem, makes a common-sense guess about what factors are worth considering, and perhaps performs a few preliminary experiments to rule out factors about which he is doubtful.

Assume that we have decided that the factors relevant to our experiment with hydrogen are temperature, pressure, and volume. In our vessel the nature and total amount of the gas remain the same because we keep it in a closed vessel. We are free, therefore, to test relationships among the three factors. If we maintain a constant temperature but increase the pressure, we discover that the volume varies inversely with the pressure. That is, if we double the pressure, the volume decreases to half its former amount. If we triple the pressure, the volume decreases to one third. This is the famous experiment performed in the seventeenth century by the Irish physicist Robert Boyle. The law he discovered, known as Boyle's law, states that if the temperature of a confined gas remains the same, the product of the volume and pressure is constant.

Next we keep the pressure constant (by leaving the same weight on the piston) but vary the temperature. We then discover that the volume increases when the gas is heated and decreases when it is cooled; and, by measuring volume and temperature, we find that volume is proportional to temperature. (This is sometimes called Charles's law, after the French scientist Jacques Charles.) We must be careful not to use either the Fahrenheit or the centigrade scale, but a scale in which zero is "absolute zero" or -273 degrees on the centigrade scale. This is the "absolute scale", or "Kelvin scale", introduced by Lord Kelvin, a nineteenth-century Scottish physicist. It is now an easy step to an experimental verification of a general law covering all three factors. Such a law is, in fact, suggested by the two laws we have already obtained, but the general law has more empirical content than the two laws taken together. This general law states that if the amount of a confined gas remains constant, the product of the pressure and volume equals the product of the temperature and R ($P \cdot V = T \cdot R$). In this equation, R is a constant that varies with the amount of gas under consideration. This general law gives the relationships among all three magnitudes and is therefore of significantly greater efficiency in making predictions than the other two laws combined. If we know the value of any two of the three variable magnitudes, we can easily predict the third.

This example of a simple experiment shows how it is possible to keep certain factors constant in order to study dependencies that hold between other factors. It also shows—and this is very important—the fruitfulness of quantitative concepts. The laws determined by this experiment presuppose the ability to measure the various magnitudes involved. If this were not so, the laws would have to be formulated in a qualitative way; such laws would be much weaker, less useful in making predictions. Without numerical scales for pressure, volume, and temperature, the most we could say about one of the magnitudes would be that it remained the same or that it increased or decreased. Thus, we could formulate Boyle's law by saying: If the temperature of a confined gas remains the same, and the pressure increases, then the volume decreases; when the pressure decreases, the volume increases. This is certainly a law. It is even similar in some ways to Boyle's law. It is much weaker than Boyle's law, however, because it does not enable us to predict specific amounts of magnitude. We can predict only that a magnitude will increase, decrease, or remain constant.

The defects of qualitative versions of the laws for gases become even more apparent if we consider the general law expressed by the equation: $P \cdot V = T \cdot R$. Let us write this in the form:

$$V = \frac{T}{P} \cdot R.$$

From this general equation, interpreted qualitatively, we can derive weak versions of Boyle's law and Charles's law. Suppose that all three magnitudes—pressure, volume, temperature—are allowed to vary simultaneously, only the quantity of gas (R) remaining constant. We find by experiment that both temperature and pressure are increasing. What can we say about the volume? In this case, we cannot say even whether it increases, decreases, or remains constant. To determine this, we would have to know the ratios by which the temperature and pressure increased. If the temperature increased by a higher ratio than the pressure, then it follows from the formula that the volume will increase. But if we cannot give numerical values to pressure and temperature, we cannot, in this case, predict anything at all about the volume.

We see, then, how little could be accomplished in the way of prediction and how crude explanations of phenomena would be if the laws of science were limited to qualitative laws. Quantitative laws are enormously superior. For such laws we must, of course, have quantitative concepts. This is the topic we shall explore in detail in Chapter 5.

Part **II**

MEASUREMENT AND QUANTITATIVE LANGUAGE

CHAPTER **5**

Three Kinds of
Concepts in Science

THE CONCEPTS OF SCIENCE, as well as those of everyday life, may be conveniently divided into three main groups: classificatory, comparative, and quantitative.

By a "classificatory concept" I mean simply a concept that places an object within a certain class. All the concepts of taxonomy in botany and zoology—the various species, families, genera, and so on—are classificatory concepts. They vary widely in the amount of information they give us about an object. For example, if I say an object is blue, or warm, or cubical, I am making relatively weak statements about the object. By placing the object in a narrower class, the information about it increases, even though it still remains relatively modest. A statement that an object is a living organism tells us much more about it than a statement that it is warm. "It is an animal", says a bit more. "It is a vertebrate", says still more. As the classes continue to narrow—mammal, dog, poodle, and so on—we have increasing amounts of, but still relatively little, information. The classificatory concepts are those most familiar to us. The earliest words a child learns—"dog", "cat", "house", "tree"—are of this kind.

More effective in conveying information are the "comparative concepts". They play a kind of intermediate role between classificatory and quantitative concepts. I think it is advisable to give attention to them because, even among scientists, the value and power of such concepts are frequently overlooked. A scientist often says: "It would certainly be desirable to introduce quantitative concepts, concepts that can be measured on a scale, into my field; unfortunately, this cannot yet be done. The field is only in its infancy. We have not yet developed techniques for measurement, so we have to restrict ourselves to a nonquantitative, a qualitative language. Perhaps in the future, when the field is more advanced, we will be able to develop a quantitative language." The scientist may be quite right in making this statement, but he is in error if he concludes that, since he has to talk in qualitative terms, he must confine his language to classificatory concepts. It is often the case that, before quantitative concepts can be introduced into a field of science, they are preceded by comparative concepts that are much more effective tools for describing, predicting, and explaining than the cruder classificatory concepts.

A classificatory concept, such as "warm" or "cool", merely places an object in a class. A comparative concept, such as "warmer" or "cooler", tells us how an object is related, in terms of more or less, to another object. Long before science developed the concept of temperature, which can be measured, it was possible to say, "This object is warmer than that object." Comparative concepts of this sort can be enormously useful. Suppose, for instance, that thirty-five men apply for a job requiring certain types of abilities and that the company has a psychologist whose task is to determine how well the applicants qualify. Classificatory judgments are, of course, better than no judgments at all. He may decide that five of the applicants have good imagination, ten of them have rather low imagination, and the rest are neither high nor low. In a similar way, he may be able to make rough classifications of the thirty-five men in terms of their manual skills, their mathematical abilities, their emotional stability, and so on. In a sense, of course, these concepts can be used as weak comparative concepts; we can say that a person with "good imagination" is higher in this ability than a person with "poor imagination". But if the psychologist can develop a comparative method that will place all thirty-five men in one rank order with respect to each ability, then we will know a great deal more about them than we knew when they were classified only in the three classes of strong, weak, and medium.

We should never underestimate the usefulness of comparative concepts, especially in fields in which the scientific method and quantitative concepts have not yet been developed. Psychology is using quantitative concepts more and more, but there are still large areas of psychology in which only comparative concepts can be applied. Anthropology has almost no quantitative concepts. It deals mostly with classificatory concepts and is much in need of empirical criteria with which to develop useful comparative concepts. In such fields, it is important to develop such concepts, which are much more powerful than the classificatory, even though it is not yet possible to make quantitative measurements.

I would like to call your attention to a monograph by Carl G. Hempel and Paul Oppenheim, *Der Typusbegriff im Lichte der neuen Logik*. It appeared in 1936, and the title means "The concept of type from the point of view of modern logic". The authors are especially concerned with psychology and related fields, in which type concepts are, as the authors emphasize, rather poor. When psychologists spend their time classifying individuals into, say, extrovert, introvert, and extrovert-introvert-intermediate, or other types, they are not really doing the best they could do. Here and there we find efforts to introduce empirical criteria, which may lead to numerical values, such as in the body typology of William Sheldon, but at the time Hempel and Oppenheim wrote their monograph, there was very little of this sort of thing. Almost every psychologist who was concerned with character, constitution, and temperament had his own type system. Hempel and Oppenheim pointed out that all these various typologies were little more than classificatory concepts. They stressed the fact that even though it would be premature to introduce measurement and quantitative concepts, it would be a great step forward if psychologists could devise workable comparative concepts.

It often happens that a comparative concept later becomes the basis for a quantitative one. A classic example is the concept "warmer", which eventually developed into "temperature". Before we go into details about the way in which empirical criteria are established for numerical concepts, however, it will be useful to see how criteria are established for comparative concepts.

To illustrate, consider the concept of weight before it was possible to give it numerical values. We have only the comparative concepts of heavier, lighter, and equal in weight. What is the empirical procedure by which we can take any pair of objects and determine how they compare in terms of these three concepts? We need only a balance scale and these two rules:

(1) If the two objects balance each other on the scale, they are of equal weight.

(2) If the objects do not balance, the object on the pan that goes down is heavier than the object on the pan that goes up.

Strictly speaking, we cannot yet say that one object has "greater weight" than the other because we have not yet introduced the quantitative concept of weight; but in actual practice, such language may be used, even though no method is yet available for assigning numerical values to the concept. A moment ago, for example, we spoke of one man as having "greater imagination" than another, although no numerical values can be assigned to imagination.

In the balance-scale illustration, as well as in all other empirical procedures for establishing comparative concepts, it is important to distinguish between those aspects of the procedure that are purely conventional and those that are not conventional because they depend on facts of nature or logical laws. To see this distinction, let us state more formally the two rules by which we define the comparative concepts of equally heavy, heavier than, and lighter than. For equality, we need a rule for defining an observable relation corresponding to equality, which I shall call "E". For the other two concepts, we need a rule for defining a relation I will call "less than" and symbolize by "L".

The relations E and L are defined by empirical procedures. We place two bodies on the two pans of a balance scale. If we observe that the scale remains in equilibrium, we say that the relation E, in respect to the property of weight, holds between the two bodies. If we observe that one pan goes up and the other down, we say that the relation L, in respect to weight, holds between the two bodies.

It may appear that we are adopting a completely conventional procedure for defining E and L, but such is not the case. Unless certain conditions are fulfilled by the two relations we chose, they cannot serve adequately as E and L. They are not, therefore, arbitrarily chosen relations. Our two relations are applied to all bodies that have weight. This set of objects is the "domain" of our comparative concepts. It must be possible to arrange all the objects of the domain into a kind of stratified structure that is sometimes called a "quasiserial arrangement". This can best be explained by using some terms from the logic of relations. The relation E, for example, must be "symmetric" (if it holds between any two bodies *a* and *b*, it must also hold between *b* and *a*). It must also be "transitive" (if it holds between *a* and *b* and between *b* and *c*, it must also hold between *a* and *c*). We can diagram this by using points to represent bodies and double arrows to indicate the relation of equality.

It is clear that if we were to choose for *E* a relation that is not symmetric, it would not be suitable for our purposes. We would have to say that one object had exactly the same weight as another but that the other object did not have the same weight as the first one. This is not, of course, the way we wish to use the term "same weight". The equilibrium of the scale is a symmetric relation. If two objects balance, they continue to balance after we have exchanged their positions on the pans. *E* must therefore be a symmetric relation. Similarly, we find that if *a* balances with *b* on the scales, and *b* balances with *c*, then *a* will balance with *c*; the relation *E* is also, therefore, transitive. If *E* is both transitive and symmetric, it must also be "reflexive"; that is, any object is equal in weight to itself. In the logic of relations, a relation that is both symmetric and transitive is called an "equivalence" relation. Our choice of the relation *E* is obviously not arbitrary. We choose as *E* the equilibrium of the scales because this relation is observed to be an equivalence relation.

The relation *L* is not symmetric; it is asymmetric. If *a* is lighter than *b*, *b* cannot be lighter than *a*. *L* is transitive: if *a* is lighter than *b* and *b* is lighter than *c*, then *a* is lighter than *c*. This transitivity of *L*, like the properties of the relation *E*, is so familiar to us that we forget we must make an empirical test to be sure it applies to the concept of weight. We place *a* and *b* on the two pans of the scale, and *a* goes down. We place *b* and *c* on the pans, and *b* goes down. If we put *a* and *c* on the pans, we expect *a* to go down. In a different world, where our laws of nature do not hold, *a* might go up. If this happened, then the relation we were testing could not be called transitive and therefore could not serve as *L*.

We can diagram the relation *L*, transitive and asymmetric, with single arrows from one point to another:

If between every pair of objects in the domain one or the other of relations E and L holds, it must be possible to arrange all the objects into the quasiserial order diagrammed in Figure 5–1. In the lowest level, stratum A,

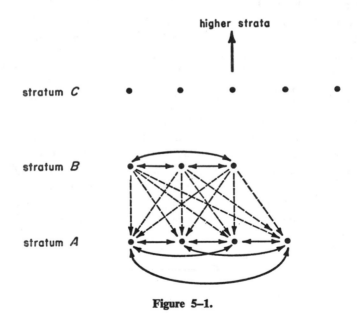

Figure 5–1.

we have all those objects that are equal in weight but lighter than all the objects not in that stratum. There may be only one such object, or there may be many thousands. Figure 5–1 shows four. In stratum B, we have another set of equally heavy objects, all related to each other by E, all heavier than objects in stratum A and lighter than all objects not in A or B. These strata continue upward, until we finally reach the stratum of the heaviest objects. Unless empirical tests show that the objects of the domain can be placed in this quasiserial arrangement, the relations E and L will not be suitable relations for defining, respectively, the comparative concepts of equal weight and less weight.

You will find all this discussed in greater detail in sections ten and eleven of Hempel's monograph, *Fundamentals of Concept Formation in Empirical Science*.[1] He says that there are four conditions that E and L must satisfy:

1. E must be an equivalence relation.
2. E and L must exclude each other. No pair of objects can be equally heavy and at the same time be so related that one is lighter than the other.
3. L must be transitive.
4. For any two objects a and b, one of the three following cases must hold. (Actually, it is sufficient to say that at least one holds. It then follows from the other conditions that exactly one will hold.)
   (a) E holds between the two objects.
   (b) L holds between a and b.
   (c) L holds between b and a.

In other words, any two objects a and b that have weight are either equal in weight, or a is lighter than b, or b is lighter than a.

If any two relations E and L fulfill these four requirements, we can say that they constitute a quasiserial order, which can be diagrammed in the stratified manner shown in Figure 5–1. By means of the equivalence relation E, we can classify all objects into equivalence classes; then, with the aid of relation L, we can place these classes into a serial order and in this way develop the whole schema of ordered strata. The point I wish to emphasize here is that comparative concepts, quite apart from the question of whether they do or do not apply to the facts of nature, are bound by a logical structure of relations.

That is not the case with classificatory concepts. In defining a class concept, we can specify any conditions we please. Of course, if we include logically contradictory conditions, such as speaking of objects that weigh three pounds and at the same instant weigh less than one pound, then we are defining a class that has, in any possible world, *no* members. Aside from this, we are free to define a class in any consistent way we wish, regardless of whether that class does or does not have members in our world. The classic example is the concept of the unicorn. We define it as an animal with the form of a horse but with a straight horn on its forehead. This is a perfectly good definition in the sense that it gives meaning to the term "unicorn". It defines a class. It is not a useful class to a zoologist, because it is empty in the empirical sense—it has no members—but this is not a question for the logician to decide.

With respect to comparative concepts, the situation is quite different. Unlike class concepts, they imply a complicated structure of logical relations. If we introduce them, we are not free to reject or modify this structure. The four requirements stated by Hempel must be fulfilled. Thus, we see that there are two ways in which the comparative concepts of science are not entirely conventional: they must apply to facts of nature, and they must conform to a logical structure of relations.

Now we come to the "quantitative concepts". Each quantitative concept has a corresponding pair of comparative concepts, which, in the development of a field of science, usually serve as a first step toward the quantitative. In the examples we have been using, the comparative concepts of less weight and of equal weight lead easily to a concept of weight that can be measured and expressed by numbers. We will discuss the nature of quantitative concepts, why they are so useful, in what fields they can be applied, and whether there are fields in which they cannot be applied. This last point is extremely important in the methodology of science, and for that reason we will go into it in greater detail. Before taking up these questions, however, I shall make some preliminary general remarks that will become clearer in the course of our discussion, but which should be stated now.

First, we must emphasize that the difference between qualitative and quantitative is not a difference in nature, but a difference in our conceptual system—in our language, we might say, if by language we mean a system of concepts. I use "language" here as logicians use it, not in the sense of English being one language and Chinese another. We have the language of physics, the language of anthropology, the language of set theory, and so on. In this sense, a language is constituted by rules for a vocabulary, rules for building sentences, rules for logical deductions from those sentences, and other rules. The kinds of concepts that occur in a scientific language are extremely important. What I want to make clear is that the difference between qualitative and quantitative is a difference between languages.

The qualitative language is restricted to predicates (for example, "grass is green"), while the quantitative language introduces what are called functor symbols, that is, symbols for functions that have numerical values. This is important, because the view is widespread, especially among philosophers, that there are two kinds of features in nature, the qualitative and the quantitative. Some philosophers maintain that modern science, because it restricts its attention more and more to quantitative features, neglects the qualitative aspects of nature and so gives an entirely distorted picture of the world. This view is entirely wrong, and we can see that it is wrong if we introduce the distinction at the proper place. When we look at nature, we cannot ask: "Are these phenomena that I see here qualitative phenomena or quantitative?" That is not the right question. If someone describes these phenomena in certain terms, defining those terms and giving us rules for their use, then we can ask: "Are these the terms of a quantitative language, or are they the terms of a prequantitative, qualitative language?"

Another important point is that conventions play a very great role in the introduction of quantitative concepts. We must not overlook this role. On the other hand, we must also be careful not to overestimate the conventional side. This is not often done, but a few philosophers have done so. Hugo Dingler in Germany is an example. He came to a completely conventionalistic view, which I regard as wrong. He said that all concepts, and even the laws of science, are a matter of convention. In my opinion, that goes too far. Poincaré also has been accused of conventionalism in this radical sense, but that, I think, is a misunderstanding of his writings. He has indeed often stressed the important role conventions play in science, but he was also well aware of the empirical components that come into play. He knew that we are not always free to make arbitrary choices in constructing a system of science; we have to accommodate our system to the facts of nature as we find them. Nature provides factors in the situation that are outside our control. Poincaré can be called a conventionalist only if all that is meant is that he was a philosopher who emphasized, more than previous philosophers, the great role of convention. He was not a radical conventionalist.

Before we take up the role of measurement in developing quantitative concepts, we should mention that there is a simpler and more basic quantitative method—the method of counting. If we had not first had the ability to count, we would be unable to measure. Counting involves nothing more than the nonnegative integers. I say "nonnegative integers" rather than "positive integers" because zero is also the result of counting if we take counting in a wide enough sense. Given a finite class—say, the class of all chairs in this room—counting is the method by which we determine the cardinal number of that class. We count the chairs—one, two, three, and so on—until we finish, on the count of twenty. Suppose we wish to count the number of pianos in a room. We look around and see no piano, so we say that the cardinal number is zero. This can be considered a degenerate case of counting. In any case, zero is an integer and can be applied to a class as its cardinal number. In such cases, we usually call it a null class.

The same counting procedure gives us the cardinal number of a finite class of consecutive events. We count the number of times that we hear thunder during a storm or the number of times a clock strikes. It is likely that this type of counting came earlier in history than the counting of classes of simultaneous things, such as chairs in a room. Indeed, it is the way in which a child first learns to count. He walks about the room, touching each individual chair while he says the number words. What he is counting, actually, is a series of events of touching. If you ask a child to count a group of trees in the distance, he finds it difficult to do so because it is hard for him to point to the trees, one by one, and carry out a form of this touching procedure. But, if he is careful in counting the events of pointing, making sure that he points to each tree once and only once, then we say that there is an isomorphism between the number of trees and the number of pointing events. If the number of these events is eight, we ascribe the same cardinal number to the class of trees in the distance.

An older child or adult may be able to count the trees without pointing. But, unless it is a small number, like three or four, which can be recognized at a glance, he concentrates his attention on first one tree, then another, and so on. The procedure is still one of counting consecutive events. That the cardinal number obtained in this way is actually the cardinal number of the class can be shown by formal proof, but we will not go into its details here. The point is that, in counting a class of objects, we actually count something else—a series of events. We then make an inference on the basis of an isomorphism (a one-one correlation between events and objects) and conclude that the cardinal number of the events is the cardinal number of the class.

A logician always finds so many complications about such simple things! Even counting, the simplest of all quantitative methods, turns out, upon analysis, to be not quite so simple as it first appears. But, once we can count, we can go on to apply rules for measurement, as explained in Chapter 6.

CHAPTER **6**

The Measurement of Quantitative Concepts

IF THE FACTS OF NATURE are to be described by quantitative concepts—concepts with numerical values —we must have procedures for arriving at those values. The simplest such procedure, as we saw in the previous chapter, is counting. In this chapter, we shall examine the more refined procedure of measurement. Counting gives only values that are expressed in integers. Measurement goes beyond this. It gives not only values that can be expressed by rational numbers (integers and fractions), but also values that can be expressed by irrational numbers. This makes it possible to apply powerful mathematical tools, such as calculus. The result is an enormous increase in the efficiency of scientific method.

The first important point that we must understand clearly is that, in order to give meaning to such terms as "length" and "temperature", we must have rules for the process of measuring. These rules are nothing other than rules that tell us how to assign a certain number to a certain body or process so we can say that this number represents the value of the magnitude for that body. As an example of how this is done, let us take the concept of temperature, together with a schema of five rules.

The rules will state the procedure by which temperature can be measured.

The first two rules of this schema are the same two rules we discussed in the last chapter as rules for defining comparative concepts. Now, however, we regard them as rules for defining a quantitative concept, which we shall call magnitude M.

Rule 1, for magnitude M, specifies an empirical relation E. The rule states that, if the relation E_M holds between objects a and b, the two objects will have equal values of the magnitude M. In symbolic form:

$$\text{If } E_M(a,b), \text{ then } M(a) = M(b).$$

Rule 2 specifies an empirical relation L_M. This rule says that, if the relation L_M holds between a and b, the value of the magnitude M will be smaller for a than for b. In symbolic form:

$$\text{If } L_M(a,b), \text{ then } M(a) < M(b).$$

Before going on to the other three rules of our schema, let us see how these two rules were first applied to the prescientific, comparative concept of temperature, then later taken over by quantitative procedures. Imagine that we are living at a time before the invention of thermometers. How do we decide that two objects are equally warm or that one is less warm than another? We touch each object with our hand. If neither feels warmer than the other (relation E), we say they are equally warm. If a feels less warm than b (relation L), we say that a is less warm than b. But these are subjective methods, very imprecise, about which it is difficult to get agreement among different observers. One person may feel that a is warmer than b; another may touch the same two objects and think the reverse is true. Memories of heat sensations are so vague that it may be impossible for a person to decide if an object feels warmer at one time than it did three hours earlier. For such reasons, subjective methods of establishing the relations "equally warm" (E) and "less warm" (L) are of little use in an empirical search for general laws. What is needed is an objective method of determining temperature—a method more precise than our sensations of heat, and one about which different persons will usually agree.

The thermometer provides just such a method. Assume that we wish to determine changes in the temperature of water in a vessel. We immerse a mercury thermometer in the water. When the water is heated, the mercury expands and rises in the tube. When the water is cooled, the mercury contracts and goes down. If a mark is placed on the tube to indicate the height of the mercury, it is so easy to see whether the mercury goes above or below the mark that two observers are not likely to disagree about it. If I observe today that the liquid is above the mark, I have no difficulty whatever in recalling that yesterday it was below the mark. I can declare, with utmost confidence, that the thermometer is registering a higher temperature today than yesterday. It is easy to see how the relations E_T and L_T, for the magnitude T (temperature), can be defined by this instrument. We simply place the thermometer in contact with body a, wait until there is no longer any change in the height of the test liquid, then mark the level of the liquid. We apply the thermometer in the same way to object b. Relation E is defined by the liquid rising to the same mark. Relation L is established between a and b if the liquid rises to a lower point, when the thermometer is applied to a, than it does when the thermometer is applied to b.

The first two rules for defining temperature (T) can be expressed symbolically as follows:

Rule 1:   If $E_T(a,b)$, then $T(a) = T(b)$.
Rule 2:   If $L_T(a,b)$, then $T(a) < T(b)$.

Note that it is not necessary, in order to establish the two relations, E and L, to have a scale of values marked on the tube. If, however, we intend to use the thermometer for assigning numerical values to T, we obviously need more than the two rules.

The three remaining rules of our schema supply the needed additional conditions. Rule 3 tells us when to assign a selected numerical value, usually zero, to the magnitude we are attempting to measure. It does this by specifying an easily recognizable, and sometimes easily reproducible, state and telling us to assign the selected numerical value to an object if it is in that state. For instance, in the centigrade temperature scale, Rule 3 assigns the value of zero to water when it is in the freezing state. Later, we will add some qualifications about the conditions under which this rule is adequate; now we shall accept it as it stands.

Rule 4, usually called the rule of the unit, assigns a second selected value of the magnitude to an object by specifying another easily recognized, easily reproducible state of that object. This second value is usually 1, but it may be any number different from the number specified by Rule 3. On the centigrade scale it is 100. It is assigned to water in the boiling state. Once the second value is assigned, a basis for defining units of temperature is available. We place the thermometer in freezing water, mark the height of the mercury, and label it zero. We then place the thermometer in boiling water, mark the height of the liquid, label it 100. We do not yet have a scale, but we do have a basis for speaking of units. If the mercury rises from the zero mark to the 100 mark, we can say that the temperature has gone up 100 degrees. If we had labeled the higher mark with the number 10, instead of 100, we would say that the temperature had risen ten degrees.

The final step is to determine the precise form of the scale. This is done by Rule 5, the most important of the five rules. It specifies the empirical conditions ED_M, under which we shall say that two differences (D) in the values of the magnitude (M) are equal. Note that we speak not of two values, but of two *differences* between two values. We want to specify empirical conditions under which we shall say that the difference between any two values of the magnitudes for a and for b is the same as the difference between two other values, say, for c and for d. This fifth rule has the following symbolic form:

If $ED_M(a,b,c,d)$, then $M(a)-M(b) = M(c)-M(d)$.

The rule tells us that if certain empirical conditions, represented by "ED_M" in the symbolic formulation, obtain for four values of the magnitude, we can say that the difference between the first two values is the same as the difference between the other two values.

In the case of temperature, the empirical conditions concern the volume of the test substance used in the thermometer, in our example, mercury. We must construct the thermometer so that when the difference between any two volumes of mercury, a and b, is equal to the difference between two other volumes, c and d, the scale will give equal differences in temperature.

If the thermometer has a centigrade scale, the procedure for fulfilling the conditions of Rule 5 is simple. The mercury is confined inside a bulb at one end of an extremely thin tube. The thinness of the tube is not essential, but it has great practical value because it makes it easy to observe extremely small changes in the volume of the mercury. The glass tube must be carefully made so that its inner diameter is uniform. As a result, equal increases in the volume of mercury can be observed as equal distances between marks along the tube. If we denote the distance between marks when the thermometer is in contact with body a and body b as "$d(a,b)$", then Rule 5 can be expressed symbolically as follows:

If $d(a,b) = d(c,d)$, then $T(a)-T(b) = T(c)-T(d)$.

Now we apply Rules 3 and 4. The thermometer is placed in freezing water, and "0" is used to mark the level of the mercury in the tube. The thermometer is placed in boiling water, and the mercury's level is marked with "100". On the basis of Rule 5, the tube can now be marked into a hundred equal space intervals between the zero and 100 marks. These intervals can be continued below zero until a point is reached at which mercury freezes. They can also be continued above 100 to the point at which mercury boils and evaporates. If two physicists construct their thermometers in this manner and agree on all the procedures specified by the five rules, they will arrive at identical results when they measure the temperature of the same object. We express this agreement by saying that the two physicists are using the same temperature scale. The five rules determine a unique scale for the magnitude to which they are applied.

How do physicists decide on the exact type of scale to use for measuring a magnitude? Their decisions are in part conventional, especially those decisions involving the choice of points in Rules 3 and 4. The unit of length, the meter, is now defined as the length, in a vacuum, of 1,656,763.83 wave lengths of a certain type of radiation from an atom of krypton 86. The unit of mass or weight, the kilogram, is based on a prototype kilogram body preserved in Paris. With respect to temperature, as measured by a centigrade scale, zero and 100 are assigned, for reasons of convenience, to freezing and boiling water. In the Fahrenheit scale and the so-called absolute, or Kelvin, scale, other states of substances are chosen for the zero and 100 points. All three scales, however, rest on essentially the same fifth-rule procedures and therefore may be considered essentially the same scale forms. A thermometer for measuring Fahrenheit temperature is constructed in exactly the same way as a thermometer for measuring centigrade temperature; they differ only in the way they are calibrated. For this reason, it is a simple matter to translate values from one scale to the other.

If two physicists adopt entirely different procedures for their fifth rule—say, one physicist correlates temperature with the expansion of the volume of mercury and another with the expansion of a bar of iron or the effect of heat on the flow of electricity through a certain device— then their scales will be quite different in form. The two scales may, of course, agree as far as Rules 3 and 4 are concerned. If each physicist has chosen the temperatures of freezing and boiling water as the two points that determine their units, then of course they will agree when they measure the temperature of freezing or boiling water. But when they apply their respective thermometers to a given pan of warm water, they are likely to get different results, and there may be no simple way of translating from one scale form to the other.

Laws based on two different scale forms will not have the same form. One scale may lead to laws that can be expressed with very simple equations. The other scale may lead to laws requiring very complex equations. It is this last point that makes the choice of fifth-rule procedures so extremely important in contrast to the more arbitrary character of Rules 3 and 4. A scientist chooses these procedures with the aim of simplifying as much as possible the basic laws of physics.

In the case of temperature, it is the absolute, or Kelvin, scale that leads to a maximum simplification of the laws of thermodynamics. The centigrade and Fahrenheit scales may be thought of as variants of the absolute scale, differing only in calibration and easily translated to the absolute scale. In early thermometers, liquids such as alcohol and mercury were used as test substances, as well as gases that were kept under constant pressure so that changes in temperature would alter their volume. It was found that no matter what substances were used, roughly identical scale forms could be established; but when more precise instruments were made, small differences could be observed. I do not mean merely that substances expand at different rates when heated, but rather that the scale form itself is somewhat different depending on whether mercury or hydrogen is used as a test substance. Eventually, scientists chose the absolute scale as the one leading to the simplest laws. The surprising fact is that this scale form was not specified by the nature of a particular test substance. It is closer to the scale of hydrogen or any other gas than to the mercury scale, but it is not exactly like any gas scale. Sometimes it is spoken of as a scale based on an "ideal gas", but that is only a manner of speaking.

In actual practice, of course, scientists continue to use thermometers containing mercury or other test liquids that have scales extremely close to the absolute scale; then they convert the temperatures based on these scales to the absolute scale by means of certain correction formulas. The absolute scale permits the formulation of thermodynamic laws in the simplest possible way, because its values express amounts of energy, rather than volume changes of various substances. Laws involving temperature would be much more complicated if any other scale form were used.

It is important to understand that we cannot really say we know what we mean by any quantitative magnitude until we have formulated rules for measuring it. It might be thought that first science develops a quantitative concept, then seeks ways of measuring it. But the quantitative concept actually develops out of the process of measuring. It was not until thermometers were invented that the concept of temperature could be given a precise meaning. Einstein stressed this point in discussions leading to the theory of relativity. He was concerned primarily with the measurement of space and time. He emphasized that we cannot know exactly what is meant by such concepts as "equality of duration", "equality of distance (in space)", "simultaneity of two events at different places", and so on, without specifying the devices and rules by which such concepts are measured.

In Chapter 5, we saw that there were both conventional and non-conventional aspects to the procedures adopted for Rules 1 and 2. A similar situation holds with respect to Rules 3, 4, and 5. There is a certain latitude of choice in deciding on procedures for these rules; to that extent, these rules are matters of convention. But they are not entirely conventional. Factual knowledge is necessary in order to decide which kinds of conventions can be carried out without coming into conflict with the facts of nature, and various logical structures must be accepted in order to avoid logical inconsistencies.

For example, we decide to take the freezing point of water as the zero point on our temperature scale because we know that the volume of mercury in our thermometer will always be the same whenever we put the bulb of the instrument into freezing water. If we found that mercury rose to one height when we used freezing water obtained from France and to a different height when we used water obtained from Denmark or that the height varied with the amount of water we were freezing, then freezing water would not be a suitable choice for applying the third rule.

A similar empirical element clearly enters into our choice of boiling water to mark the 100 point. It is a fact of nature, not a matter of convention, that the temperature of all boiling water is the same. (We assume that we have already established Rules 1 and 2, so that we have a way to measure equality of temperature.) But here we must introduce a qualification. The temperature of boiling water is the same in the same locality, but on a high mountain, where air pressure is less, it boils at a slightly lower temperature than it does at the foot of the mountain.

In order to use the boiling point of water to meet the demands of the fourth rule, we must either add that we have to use boiling water at a certain altitude or apply a correction factor if it is not at that altitude. Strictly speaking, even at the specified altitude we must make sure, by means of a barometer, that we have a certain specified air pressure, or a correction would have to be applied there also. These corrections depend on empirical facts. They are not conventional, arbitrarily introduced factors.

In finding empirical criteria for applying Rule 5, which determines the form of our scale, we seek a form that will give the simplest possible laws. Here again, a nonconventional aspect enters into the choice of the rule, because the facts of nature determine the laws we seek to simplify. And finally, the use of numbers as values on our scale implies a structure of logical relations that are not conventional because we cannot abandon them without becoming entangled in logical contradictions.

CHAPTER 7

Extensive Magnitudes

THE MEASUREMENT OF temperature requires, as we learned in Chapter 6, a schema of five rules. Are there concepts in physics that can be measured by the use of simpler schemas? Yes, a large number of magnitudes, called "extensive magnitudes", are measurable with the aid of three-rule schemas.

Three-rule schemas apply to situations in which two things can be combined or joined in some way to produce a new thing, and the value of a magnitude M for this new thing will be the sum of the values of M for the two things that were joined. Weight, for example, is an extensive magnitude. If we place together a five-pound object and a two-pound object, the weight of the combined objects will be seven pounds. Temperature is not such a magnitude. There is no simple operation by which we can take an object with, say, a temperature of 60 degrees, combine it with an object that has a temperature of 40 degrees, and produce a new object with a temperature of 100 degrees.

The operations by which extensive magnitudes are combined vary enormously from magnitude to magnitude. In the simplest cases, the operation is merely the putting together of two bodies, gluing them together, or tying them, or perhaps just placing them side by side, like two weights on the same pan of a balance scale. Daily life abounds with examples. The width of a row of books on a shelf is the sum of the individual widths of the books. We take down a book and read ten pages. Later in the day we read ten more pages. Altogether, we have read twenty pages. After partially filling a bathtub, we discover that the water is too hot, so we add some cold water. The total volume of water in the tub will be the sum of the amounts of hot and cold water that came through the faucets. The exact procedure for combining things with respect to a certain extensive magnitude is often not explicitly stated. This is a risky practice and can cause great confusion and misunderstanding. Because there are so many different ways things can be combined, it is important not to assume that the method of combining is understood. It should be explicitly stated and clearly defined. Once this has been done, the magnitude can be measured by employing a three-rule schema.

The first rule lays down what is called the principle of addition, or "additivity". This states that, when a combined object is made out of two components, the value of the magnitude for that object is the arithmetical sum of the values of the magnitude for the two components. Any magnitude that conforms to this rule is called an "additive magnitude". Weight is a familiar example. The joining operation in that case is simply the placing together of two objects and weighing them as a single object. We put object a on the scale and observe its weight. We replace it with object b and note its weight. Then we put both objects on the scale. This new object, which is nothing more than a and b taken together, will, of course, have a weight that is the arithmetical sum of the weights of a and b.

If this is the first time the reader has encountered this rule, he may think it strange that we even mention such a trivial rule. But in the logical analysis of scientific method, we must make everything explicit, including matters that the man on the street takes for granted and seldom puts into words. Naturally, no one would think that, if a stone of five pounds were placed on a scale alongside a stone of seven pounds, the scale would register a total weight of 70 pounds or of three pounds. We take for granted that the combined weight will be twelve pounds. It is conceivable, however, that in some other world the magnitude of weight might not behave in so convenient an additive fashion. We must, therefore, make the additivity of weight explicit by introducing this additive rule: if two bodies are joined and weighed as though they were one, the total weight will be the arithmetical sum of the component weights.

Similar rules must be introduced for every extensive magnitude. Spatial length is another familiar example. One body has a straight edge *a*. Another body has a straight edge *b*. We place the two together so that the two edges are end to end and lying on one straight line. This new physical entity—the straight line formed by combining *a* and *b*—will have a length that is the sum of the lengths of *a* and *b*.

Early formulations of the additive rule for length were frequently quite unsatisfactory. For example, some authors said that if two line segments, *a* and *b*, were added, the length of the new segment was obtained by adding the length of *a* and the length of *b*. This is an extremely poor way to formulate the rule, because in the same sentence the word "add" is used in two entirely different ways. First it is used in the sense of joining two physical objects by putting them together in a specified way, and then it is used in the sense of the arithmetical operation of addition. These authors were apparently unaware that the two concepts are different, because when they proceeded to symbolize the rule, they wrote it this way:

$$L(a + b) = L(a) + L(b).$$

Some authors, whom I otherwise admire, were guilty of this clumsy formulation, a formulation that carries over into symbols the same double use of the word "add". The second symbol "+" designates an arithmetical operation, but the first "+" is not an arithmetical operation at all. You cannot arithmetically add two lines. What you add are not the lines, but numbers that represent the lengths of the lines. The lines are not numbers; they are configurations in physical space. I have always stressed that a distinction must be made between arithmetical addition and the kind of addition that constitutes the physical operation of combining. It helps us to keep this distinction in mind if we follow Hempel (who has written at length about extensive magnitudes) in introducing a special symbol, a small circle, "∘", for the physical operation of joining. This provides a much more satisfactory way of symbolizing the additive rule for length:

$$L(a \circ b) = L(a) + L(b).$$

The combining of lengths can be diagrammed:

$$
\begin{array}{cc}
a & b \\
\overline{\hspace{1em}} & \overline{\hspace{1em}} \\
L(a) & L(b) \\
\end{array}
$$

$$L(a \circ b) \qquad [\text{not } "L(a + b)"]$$

Although in the case of weight it does not matter exactly how the two bodies are placed together on the scale, it *does* matter in the case of length. Suppose that two line segments are placed like this:

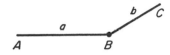

They are end to end, but not in a straight line. The distance between points *A* and *C* is not the sum of the lengths of *a* and *b*. We must always be careful, therefore, to specify exactly what we mean by the operation of joining.

We can now symbolize the general principle of additivity, with respect to any extensive magnitude *M,* by writing:

$$M(a \circ b) = M(a) + M(b).$$

In this statement, the symbol "∘" indicates a specified procedure for joining *a* and *b*. It will be best if we call this the second rule of our three-rule schema, rather than the first rule. The first rule, which is simpler, is the rule of equality. It is the same as the first rule of the five-rule schema for measuring temperature. It specifies the procedure by which we define equality of magnitude. In the case of weight, we say that two bodies have the same weight if, when they are placed on two sides of the balance scale, the scale remains in equilibrium.

The third rule corresponds to Rule 4 of the schema for temperature. It specifies the unit of value for the magnitude. This is usually done by choosing an object or a natural process that can be easily reproduced, then defining the unit of value in terms of that object or process. I mentioned earlier two examples: the meter, based on so many wave lengths of a certain type of light, and the kilogram, based on an international prototype in Paris. The meter and kilogram are the standard units of length and weight in the metric system of measurements.

To summarize, our schema for the measurement of any extensive magnitude consists of the following three rules:

1. The rule of equality.
2. The rule of additivity.
3. The unit rule.

Since this is a simpler schema than the previously discussed five-rule schema, why isn't it always used? The answer, of course, is that for many magnitudes there is no operation of joining to provide a basis for the additive principle. We have already seen that temperature is a non-additive magnitude. The pitch of sound and the hardness of bodies are two other examples. With respect to these magnitudes, we cannot find a joining operation that is additive. Such magnitudes are called "nonextensive" or "intensive magnitudes". There are, however, a large number of additive magnitudes in physics, and, with respect to all of them, the above three-fold schema provides an adequate basis for measurement.

Many scientists and philosophers of science take the terms "extensive magnitudes" and "additive magnitudes" as synonymous, but there are some authors who distinguish between them. If we make such a distinction, it should be done in this way. We call a magnitude extensive if we can think of an operation that seems to be a natural operation of joining and for which a scale can be devised. If we then discover that, with respect to the chosen scale and the chosen operation, the additive principle holds, we call it an additive magnitude as well as an extensive one. We might say it is an additive-extensive magnitude. If, however, the additive principle does not hold, we call it a nonadditive-extensive magnitude.

Almost all the extensive magnitudes of physics are additive, but there are some exceptions. A notable example is relative velocity in the special theory of relativity. In classical physics, relative velocities along a straight line are additive in the following sense. If bodies A, B, C move on a straight line in the same direction, and the velocity of B relative to A is V_1 and the velocity of C relative to B is V_2, then in classical physics the velocity V_3 of C relative to A was taken simply as equal to $V_1 + V_2$. If you walk forward along the central aisle of a plane flying due west, what is your westward velocity relative to the ground? Before relativity theory, this would have been answered simply by adding the plane's velocity to your forward walking speed inside the plane. Today we know that relative velocities are not additive; a special formula must be used in which the velocity of light is one of the terms. When velocities are small in relation to that of light, they can be handled as if they were additive; but when velocities are extremely large, this formula must be used, where c is the velocity of light:

$$V_3 = \frac{V_1 + V_2}{1 + \dfrac{V_1 V_2}{c^2}}$$

Imagine, for example, that spaceship B is moving in a straight path and passing the planet A with a relative velocity V_1. Spaceship C, traveling in the same direction, passes spaceship B with a velocity V_2 (relative to B). What is the relative velocity, V_3, of spaceship C, with respect to the planet A? If the velocities V_1 and V_2 of the spaceships are small, then the value of the fraction to be added to 1, below the line on the right of the formula, will be so small that it can be ignored. We then obtain V_3 simply by adding V_1 and V_2. But if the spaceships are traveling at very great velocities, the speed of light, c, becomes a factor to be taken into consideration. V_3 will deviate significantly from the simple sum of V_1 and V_2. If you study the formula, you will see that no matter how closely the relative velocities of the spaceships approach the speed of light, the sum of the two velocities cannot exceed the speed of light. We conclude, therefore, that relative velocity in the special theory of relativity is extensive (because a joining operation can be specified) but not additive.

Other examples of extensive-nonadditive magnitudes are the trigonometric functions of angles. Suppose you have an angle α between the straight edges L_1 and L_2 of a piece of sheet metal A (see Figure 7–1).

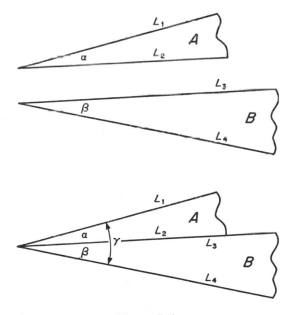

Figure 7–1.

Another piece of sheet metal, B, has the angle β between the edges L_3 and L_4. We now join the two angles by placing them together on the top of a table so that their vertices coincide, and L_2 of A coincides with part of L_3 of B. Angle γ between L_1 and L_4 is clearly the result of joining angles α and β. We can say, therefore, that when angles are joined in this manner and measured in the customary way, their values are additive. Angle γ has a value that is the sum of the values of α and β. But their values are *not* additive if we take as our magnitude one of the trigonometric functions, such as the sine, of each angle. If we wish, we can call the sine magnitude extensive (because we have a joining operation) but not additive. On the other hand, we may decide that we do not wish to call the sine extensive because the joining operation does not really join the sines. It joins the angles, but this is not quite the same as putting together the sines. From this second point of view, the sine is not extensive.

The criterion we suggested for deciding whether a magnitude is or is not extensive is, we find, not exact. As you recall, we said that if we can think of an operation that *seems* to us a natural operation of joining, with respect to the given magnitude, then we call that operation extensive. One person may say that for him the operation of placing two angles side by side is a completely natural way of joining sines. For him, then, the sine is a nonadditive-extensive magnitude. Someone else might say that it is a perfectly good operation for joining angles, but not for joining sines. For that person, the sine is not extensive. In other words, there are boundary cases in which whether to call a magnitude extensive or not is a subjective matter. Since these cases of extensive but nonadditive magnitudes are relatively rare and even questionable (questionable because we might not be willing to accept the proposed operation as one of legitimate joining), it is quite understandable that many authors use "extensive" and "additive" as synonymous terms. There is no need to criticize such usage. For those authors, "extensive" is applied to a magnitude only if there is a joining operation with respect to which the additivity principle holds, as it holds for length, weight, and many of the common magnitudes of physics.

Some remarks about the measurement of temporal intervals and spatial lengths are now in order, because, in a certain sense, these two magnitudes are basic in physics. Once we can measure them, many other magnitudes can be defined. It may not be possible to define those other magnitudes explicitly, but at least they can be introduced by operational rules that make use of the concepts of distance in space or time. You remember, for example, that in the rules for measuring temperature we made use of the concept of volume of the mercury and the length of a mercury column in a tube. In that instance, we presupposed that we already knew how to measure length. In measuring many other magnitudes of physics, similar reference is made to the measurements of length in space and duration in time. In this sense, length and duration may be regarded as primary magnitudes. Chapters 8 and 9 will discuss the procedures by which time and space are measured.

CHAPTER 8

Time

WHAT SORT OF joining operation can be used for combining time intervals? We are immediately faced with a grave difficulty. We cannot manipulate time intervals in the way we can manipulate space intervals, or, more accurately, edges of solid bodies representing space intervals. There are no hard edges of time that can be put together to form a straight line.

Consider these two intervals: the length of a certain war from the first shot to the last and the duration of a certain thunderstorm from the first thunderclap to the last. How can we join those two durations? We have two separate events, each with a certain length of time, but there is no way to bring them together. Of course, if two events are already together in time, we can recognize that fact, but we cannot shift events around as we can shift the edges of physical objects.

The best we can do is to represent the two time intervals on a conceptual scale. Suppose we have one event a that ran from time point A to time point B and a second event b that ran from time point B to time point C. (See Figure 8–1.) The initial point of b is the same as the terminal point of a, so the two events are adjacent in time. We did not

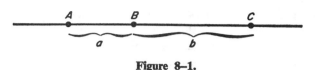

Figure 8–1.

push them into this position—that is how they occurred. The length of time from point A to point C can now be regarded as the result of combining a and b, not in the physical way lengths are combined, but in a conceptual way, that is, by the way we look at this situation. The conceptual operation, symbolized by "$\circ$", allows us to formulate the following rule of additivity for the measurement of temporal length T:

$$T(a \circ b) = T(a) + T(b).$$

In other words, if we have two events, one beginning just as the other ends, then the length of the total event will be the arithmetical sum of the lengths of the two events. This is not so powerful as the additivity rule for spatial lengths because we can apply it only to events that happen to be adjacent in time, not to any pair of events. Later, after we have developed a three-rule schema for measuring time, we will be able to measure the combined lengths of nonadjacent events. Now we seek only a joining operation that will furnish the basis for an additivity rule. We find this operation in the occurrence of events adjacent in time.

To complete our schema, we need two more rules: a rule of equality and a rule that will define a unit. Both rules are usually based on some type of periodic process: a swinging pendulum, the rotation of the earth, and so on. Every clock is simply an instrument for creating a periodic process. In some clocks, this is accomplished by a pendulum, in others, by a balance wheel. The sun dial measures time by the periodic movement of the sun across the sky. For thousands of years, scientists based their units of time on the length of the day, that is, on the periodic rotation of the earth. But, because the rate of the earth's spin is changing slightly, in 1956 an international agreement was reached to base units of time on the periodic movement of the earth around the sun in one particular year. The second was defined as 1/31,556,925.9747 of the year 1900. This was abandoned in 1964 so that still greater precision could be obtained by basing the second on the periodic vibration rate of the cesium atom. This concept of "periodicity", so essential in defining time units, must be fully understood before we consider how a rule of equality and a unit rule can be based on it.

We must first clearly distinguish the two meanings of "periodicity", one weak, the other strong. In the weak sense, a process is periodic simply if it recurs again and again and again. A pulse beat is periodic. A swinging pendulum is periodic. But so also, in a weak sense, is the exit of Mr. Smith from his house. It occurs again and again, hundreds of times, during Mr. Smith's lifetime. It is clearly periodic in the weak sense of being repeated. Sometimes periodic means that a total cycle of different phases is repeated in the same cyclic order. A pendulum, for example, swings from its lowest point up to its highest point on the right, back down past its lowest point, up to its highest point on the left, and back to the lowest point; then the entire cycle is repeated. Not just one event, but a sequence of events recurs. This is not, however, necessary in order to call a process periodic. It is sufficient if one phase of the process continues to repeat. Such a process is periodic in the weak sense.

Frequently, when someone says that a process is periodic, he means it in a much stronger sense: that in addition to being weakly periodic, it is also true that the intervals between successive occurrences of a certain phase are equal. With respect to Mr. Smith's exits from his home, this condition is obviously not fulfilled. On some days he may stay in the house many hours. On other days he may leave the house several times within an hour. In contrast, the movements of the balance wheel of a well-constructed clock are periodic in the strong sense. There is clearly an enormous difference between the two types of periodicity.

Which type of periodicity should we take as the basis for measuring time? At first we are inclined to answer that obviously we must choose a process that is periodic in the strong sense. We cannot base the measurement of time on the exit of Mr. Smith from his house because it is too irregular. We cannot even base it on a pulse because, although a pulse comes much closer to being periodic in the strong sense than Mr. Smith's exit, it is still not regular enough. If one has been running hard or has a high fever, his pulse beats faster than at other times. What we need is a process that is periodic in the strongest possible way.

But there is something wrong with this reasoning. We cannot know that a process is periodic in the strong sense unless we already have a method for determining equal intervals of time! It is precisely such a method that we are trying to establish by our rules. How can we escape this vicious circle? We can escape only by dispensing entirely with the

requirement of periodicity in the strong sense. We are forced to abandon it, because we do not yet have a basis for recognizing it. We are in
the position of a naïve physicist approaching the problem of measuring
time without even the advantage of prescientific notions of equal time
intervals. Having no basis whatever for time measurement, he is seeking an observable periodic process in nature that will furnish such a
basis. Since he has no way of measuring time intervals, he has no way
of discovering whether a particular process is or is not periodic in the
strong sense.

This is what we must do. First, we find a process that is periodic
in the weak sense. (It may also be periodic in the strong sense, but that
is something we cannot yet know.) Then we take as our joining operation two intervals of time that are consecutive in the sense that one begins just as the other ends, and we affirm, as our rule of additivity, that
the length of the total interval is the arithmetical sum of the lengths of
the two component intervals. We can then apply this rule to the chosen
periodic process.

To complete our schema, we must find rules for equality and for
the unit. The duration of any one of the periods of the chosen process
can serve as our unit of time. In Figure 8–2, these periods are dia-

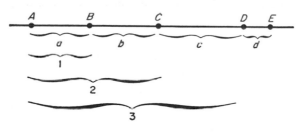

Figure 8–2.

grammed as the lengths a, b, c, d . . . between the time points $A, B,$
C, D, E . . . We say that each of these segments has a length of one
unit. Someone may object: "But period b took much longer than period
a." We reply: "We do not know what you mean by 'longer'. We are trying to lay down rules for the measurement of time so that we will be able
to give meaning to the term 'longer'."

Now that we have specified our unit (it is simply the length of each
period of the selected process), our additive rule provides us with a
basis for measuring time lengths. This rule tells us that the time interval

from point A to point C is 2, from point A to point D is 3, and so on. We can now measure any interval of time, even though we are basing our procedure on a weakly periodic process. We simply count the number of times that our unit period occurs while the event we wish to measure is taking place. That number will be the length of the event. The rule for equality is obvious. It says that two time intervals (which may possibly be widely separated in time) are equal if both contain the same number of elementary periods of the periodic process. This completes our three-rule schema. We have a rule of equality, a rule of additivity, and a unit rule. On the basis of this schema we have a method for measuring time.

There may be objections. Can such a schema really be based on any weakly periodic process? For example, can it be based on Mr. Smith's exits from his home? The surprising answer is yes, although, as I will explain in a moment, the laws of physics are much simpler if we choose certain other processes. The important point to understand now is that once we have established a schema for measuring time, even though based on a process as irregular as Mr. Smith's exits, we have acquired a means for determining whether one periodic process is equivalent to another.

Assume that we have adopted as our basis for measuring time the periodic process P. We can now compare P with another weakly periodic process P' to see if they are "equivalent". Suppose, for example, that P, our chosen periodic process, is the swing of a certain short pendulum. We wish to compare it with P', the swing of a longer pendulum. In view of the fact that the periods of the two pendulums are not equal, how do we compare the two? We do it by counting the swings of both pendulums during a longer time interval. We may discover that ten swings of the short pendulum coincide with six swings of the long. This occurs every time we repeat the test. We are not yet able to deal with fractions of periods, so our comparison must be in terms of integral numbers of swings. We may observe, however, that the coincidence is not exact. After ten swings of the short pendulum, the long one has already started on its seventh swing. We refine our comparison by taking a longer time interval, such as one hundred periods of the short pendulum. We discover, each time the test is repeated, that during this interval the long pendulum has sixty-two periods. In this way, we can sharpen our comparison as much as we please. If we find that a certain number of periods of process P always match a certain number of periods of process P', we say that the two periodicities are equivalent.

It is a fact of nature that there is a very large class of periodic processes that are equivalent to each other in this sense. This is not something we could know a priori. We discover it by observing the world. We cannot say that these equivalent processes are strongly periodic, but we can compare any two of them and find that they are equivalent. All swinging pendulums belong to this class, as do the movements of balance wheels in clocks and watches, the apparent movement of the sun across the sky, and so on. We find in nature an enormous class of processes in which any two processes prove to be equivalent when we compare them in the manner explained in the previous paragraph. As far as we know, there is only *one* large class of this kind.

What happens if we decide to base our time scale on a periodic process that does not belong to this large class of equivalent processes, such as the beat of a pulse? The results will be somewhat strange, but we want to emphasize that the choice of a pulse beat for the basis of time measurement will not lead to any logical contradiction. There is no sense in which it is "false" to measure time on such a basis.

Imagine that we are living in a very early phase of the development of concepts for measurement. We possess no instrument for measuring time, such as a watch, so we have no way of determining how our pulse beat may vary under different physiological circumstances. We are seeking, for the first time, to develop operational rules for measuring time, and we decide to use my pulse beat as the basis of measurement.

As soon as we compare my pulse beat with other periodic processes in nature, we find that all sorts of processes, which we might have thought uniform, turn out not to be. For example, we discover that it takes the sun so many pulse beats of time to cross the sky on days when I am feeling good. But on days when I have a fever, it takes the sun much longer to make the trip. We find this strange, but there is nothing logically contradictory in our description of the entire world on this basis. We cannot say that the pendulum is the "right" choice as the basis for our time unit and my pulse beat the "wrong" choice. No right or wrong is involved here because there is no logical contradiction in either case. It is merely a choice between a simple and a complex description of the world.

If we base time on my pulse, we have to say that all sorts of periodic processes in nature have time intervals that vary, depending on what I am doing or how I feel. If I run fast for a time and then stop running and measure these natural processes by means of my pulse, I find that while I am running and for a short time thereafter, things in the world slow down. In a few minutes, they return to normal again. You must remember that we are supposing ourselves to be in an age before we have acquired any knowledge of the laws of nature. We have no physics textbooks to tell us that this or that process is uniform. In our primitive system of physics, the revolution of the earth, the swinging of pendulums, and so on, are very irregular. They have one speed when I am well, another when I have a fever.

We thus have a genuine choice to make here. It is not a choice between a right and wrong measuring procedure, but a choice based on simplicity. We find that if we choose the pendulum as our basis of time, the resulting system of physical laws will be enormously simpler than if we choose my pulse beat. It is complicated enough if we use my pulse beat, but of course it would be much worse if we chose the exits of Mr. Smith from his house, unless our Mr. Smith were like Immanuel Kant, who is said to have come out of his house at so precisely the same time every morning that people in the community adjusted their clocks by his appearance on the street. But no ordinary mortal's movements would be a suitable basis for time measurement.

By "suitable" I mean, of course, convenient in the sense of leading to simple laws. When we base our measurement of time on the swing of a pendulum, we find that the entire universe behaves with a great regularity and can be described with laws of great simplicity. The reader may not have found those laws simple when he learned physics, but they are simple in the relative sense that they would be much more complicated if we adopted the pulse beat as our time unit. Physicists are constantly expressing surprise at the simplicity of new laws. When Einstein discovered his general principle of relativity, he voiced amazement at the fact that such a relatively simple principle governed all the phenomena to which it applied. This simplicity would disappear if we based our system of time measurement on a process that did not belong to the very large class of mutually equivalent processes.

My pulse beat belongs, in contrast, to an exceedingly small class of equivalent processes. The only other members are probably events of my own body that are physiologically connected with the heart beat. The pulse in my left wrist is equivalent to the pulse in my right wrist. But aside from events that have to do with my heart, it would be difficult to find another process anywhere in nature to which my pulse would be equivalent. We thus have here an extremely small class of equivalent processes as compared to the one very comprehensive class that includes the motions of the planets, the swinging of pendulums, and so on. It is advisable, therefore, to choose a process to serve as a basis for time measurement from this large class.

It does not matter much which one of this class we take, since we are not yet concerned with great precision of measurement. Once we make the choice, we can say that the process we have chosen is periodic in the strong sense. This is, of course, merely a matter of definition. But now the other processes that are equivalent to it are strongly periodic in a way that is not trivial, not merely a matter of definition. We make empirical tests and find by observation that they are strongly periodic in the sense that they exhibit great uniformity in their time intervals. As a result, we are able to describe the processes of nature in a relatively simple way. This is such an important point that I emphasize it by repeating it many times. Our choice of a process as a basis for the measurement of time is not a matter of right or wrong. Any choice is logically possible. Any choice will lead to a consistent set of natural laws. But if we base our measurement of time on such processes as the swinging of a pendulum, we find that it leads to a much simpler physics than if we use certain other processes.

Historically, our physiological sense of time, our intuitive feeling of regularity, undoubtedly entered into early choices of what processes to adopt as a basis for time measurement. The sun seems to rise and set regularly, so sun dials became a convenient way to measure time—much more convenient, for example, than the movements of clouds. In a similar way, early cultures found it convenient to base clocks on the time of running sand, or running water, or other processes that were roughly equivalent to the movement of the sun. But the basic point remains: a choice is made in terms of convenience and simplicity.

CHAPTER 9

Length

LET US TURN, now, from the concept of time to the other basic concept of physics, length, and examine it more closely than we have before. You will recall that in Chapter 7 we saw that length was an extensive magnitude, measurable by means of a three-fold schema. Rule 1 defines equality: a segment marked on one straight edge has equal length with another segment, marked on another straight edge, if the endpoints of the two segments can be brought into simultaneous coincidence with each other. Rule 2 defines additivity: if we join two edges in a straight line, their total length is the sum of their separate lengths. Rule 3 defines the unit: we choose a rod with a straight edge, mark two points on this edge, and choose the segment between those two points as our unit of length.

On the basis of these three rules we can now apply the customary procedure of measuring. Suppose we wish to measure the length of a long edge c, say, the edge of a fence. We have a measuring rod on which our unit of length a is marked by its endpoints A and B. We place the rod alongside c, in the position a_1 (see Figure 9–1), so that A coincides with one endpoint C_0 of c. On edge c we mark the point C_1 that coincides with end B of our rod. Then we move the rod a into the ad-

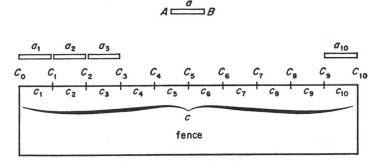

Figure 9–1.

jacent position a_2 and mark point C_2 on c, and so on, until we reach the other end of c. Suppose that the tenth position a_{10} of the rod is such that its endpoint B coincides roughly with endpoint C_{10} of c. Let c_1, c_2, . . . , c_{10} be the marked segments of c. We have by Rule 3:

$$L(a) = L(a_1) = L(a_2) = \ldots = L(a_{10}) = 1.$$

Therefore, by Rule 1, of equality:

$$L(c_1) = 1, L(c_2) = 1, \ldots L(c_{10}) = 1.$$

By Rule 2, of additivity:

$$L(c_1 \circ c_2) = 2, L(c_1 \circ c_2 \circ c_3) = 3. \ldots$$

Therefore:

$$L(c) = L(c_1 \circ c_2 \circ \ldots \circ c_{10}) = 10.$$

This procedure, the basic procedure for measuring length, yields only integers as values of the measured length. The obvious refinement is made by dividing the unit of length into n equal parts. (The inch is traditionally divided in a binary way: first into two parts, then into four, eight, and so on. The meter is divided decimally: first into ten parts, then into a hundred, and so on.) In this way, we are able to construct, by trial and error, an auxiliary measuring rod with a marked segment of length d, such that d can be put into n adjacent positions, $d_1, d_2, \ldots, d_n$, along the unit edge a (see Figure 9–2). We can now say that:

$$n \times L(d) = L(a) = 1$$

Therefore:
$$L(d) = \frac{1}{n}$$

With these partial segments marked on a, we can now measure the length of a given edge more precisely. When we remeasure the length of fence c, in the previous example, the length may now come out, not as 10, but more precisely as 10.2. In this way, fractions are introduced into measurements. We are no longer limited to integers. A measured value can be any positive rational number.

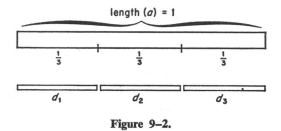

Figure 9–2.

It is important to understand that, by making these refinements in measurement, we can introduce smaller and smaller fractions, but we can never arrive at numbers that are not rational. On the other hand, the class of the possible values of a magnitude in physics is usually regarded as containing all real numbers (or all real numbers of a specified interval) which includes the irrational numbers as well as the rational. These irrational numbers, however, are introduced at a stage later than that of measurement. Direct measurement can give only values expressed as rational numbers. But when we formulate laws, and make calculations with the help of those laws, then irrational numbers enter the picture. They are introduced in a theoretical context, not in the context of direct measurement.

To make this clearer, consider the theorem of Pythagoras that states that the square of the hypotenuse of a right triangle equals the sum of the squares of the other two sides. This is a theorem in mathematical geometry, but, when we apply it to physical segments, it becomes a law of physics also. Suppose that we cut from a wooden board a square with a side of unit length. The Pythagorean theorem tells us that the length of this square's diagonal (see Figure 9–3) is the square root of 2. The square root of 2 is an irrational number. It cannot strictly be measured with a ruler based on our unit of measurement, no matter how small we mark fractional subdivisions. However, when we calculate the length of the diagonal, using the Pythagorean theorem, we obtain, indirectly, an irrational number. In a similar way, if we measure the diameter of a circular wooden disk and find it to be 1, we calculate the length of the disk's perimeter as the irrational number pi.

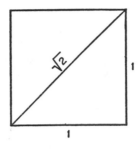

Figure 9–3.

Because irrational numbers are always the results of calculations, never the result of direct measurement, might it not be possible in physics to abandon irrational numbers altogether and work only with the rational numbers? That is certainly possible, but it would be a revolutionary change. We would, for instance, no longer be able to work with differential equations, because such equations require the continuum of real numbers. Physicists have not yet found important enough reasons for making such a change. It is true, however, that in quantum physics a trend toward discreteness is beginning. The electric charge, for example, is measured only in amounts that are multiples of a minimum electrical charge. If we take this minimum charge as the unit, all values of electrical charges are integers. Quantum mechanics is not yet completely discrete, but so much of it is discrete that some physicists are beginning to speculate on the possibility that all physical magnitudes, including those of space and time, are discrete. This is only a speculation, although it is a most interesting one.

What sort of laws would be possible in such a physics? There would probably be a minimum value for each magnitude, and all larger values would be expressed as multiples of this basic value. It has been suggested that the minimum value for length be called a "hodon", and the minimum value for time be called a "chronon". Discrete time would consist of inconceivably minute jumps, like the motion of the hand of an electric clock as it jumps from one second to the next. No physical event could occur within any interval between jumps.

Discrete space might consist of points of the sort shown in Figure 9–4. The connecting lines in the diagram indicate which points are "neighboring points" (for example, B and C are neighbors, B and F are not). In the customary geometry of continuity, we would say that there

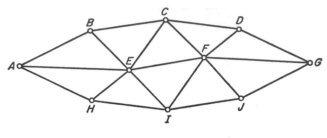

Figure 9–4.

is an infinity of points between *B* and *C*, but in discrete geometry, if physics adopted this view of space, we would have to say that there are *no* intermediate points between *B* and *C*. No physical phenomenon, of any sort, can have a position "between" *B* and *C*. An electron, for instance, would have to be at one of the points on the network, never anywhere else on the diagram. Length would be defined as the minimum length of a path connecting two points. We could stipulate that the distance between any two neighboring points is 1. Then the length of path *ABCDG* would be 4, that of *AEFG* would be 3. We would say that the distance from *A* to *G* is 3, because that is the length of the shortest path from *A* to *C*. Every length would be expressed as an integer. No actual system of this kind has been constructed for physics, although many suggestive hints have been made. Some physicists have even speculated on the size of these minimum magnitudes.

At some future time, when much more is known about space and time and the other magnitudes of physics, we may find that all of them are discrete. The laws of physics would then deal solely with integers. They would, of course, be integers of stupendous size. In each millimeter of length, for example, there would be billions of the minimum unit. The values assumed by a magnitude would be so close to each other that, in practice, we could proceed as if we had a continuum of real numbers. Practically, physicists would probably continue to use calculus and formulate laws as differential equations, just as before. The most we can say now is that some features of physics would be simplified by adopting discrete scales, whereas others would become more complicated. Our observations can never decide whether a value must be expressed as a rational or irrational number, so the question here is entirely one of convenience—will a discrete or a continuous number scale be the most useful for formulating certain physical laws?

In our description of how lengths are measured, one extremely important question has not yet been considered—what kind of body shall we take as our standard measuring rod? For everyday purposes, it would be sufficient to take an iron rod, or even a wooden rod, because here it is not necessary to measure lengths with great precision. But, if we are seeking greater accuracy, we see at once that we are up against a difficulty similar to the one we encountered with respect to periodicity.

We had, you recall, the apparent problem of basing our time unit on a periodic process with equal periods. Here we have the analogous problem of basing our unit of length on a "rigid body". We are inclined to think that we need a body that will always remain exactly the same length, just as before we needed a periodic process with time intervals that were always the same. Obviously, we think, we do not want to base our unit of length on a rubber rod or on one made of wax, which is easily deformed. We assume that we need a rigid rod, one that will not alter its shape or size. Perhaps we define "rigidity" this way: a rod is rigid if the distance between any two points marked on the rod remains constant in the course of time.

But exactly what do we mean by "remains constant"? To explain, we would have to introduce the concept of length. Unless we have a concept of length and a means of measuring it, what would it mean to say that the distance between two points on a rod does, in fact, remain constant? And if we cannot determine this, how can we define rigidity? We are thus trapped in the same sort of circularity in which we found ourselves trapped when we sought a way to identify a strongly periodic process before we had developed a system of time measurement. Once again, how do we escape the vicious circle?

The way out is similar to the way by which we escaped from circularity in measuring time: the use of a relative instead of an absolute concept. We can, without circularity, define a concept of "relative rigidity" of one body with respect to another. Take one body M and another body M'. For simplicity's sake, we assume that each has a straight edge. We can place the edges together and compare points marked along them. (See Figure 9–5.)

Consider a pair of points A, B on M that determine the segment a. Similarly, on M' a pair of points, A', B', determine the segment a'. We say that segment a is congruent with segment a' if, whenever the two edges are put alongside each other, so that point A coincides with point A', point B coincides with B'. This is our operational procedure for de-

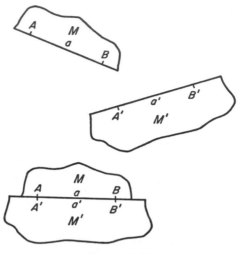

Figure 9–5.

ciding that segments *a* and *a′* are congruent. We find that, whenever we make this test, the point pairs coincide, so we conclude that, if we repeated the experiment at any time in the future, the result would probably be the same. In addition, suppose that *every* segment marked in this way on *M* is found to be congruent, every time a test is made, with its corresponding segment marked on *M′*. We then say that *M* and *M′* are *rigid with respect to each other*.

It is important to realize that no circularity is involved here. We cannot and do not speak of absolute rigidity of *M*; we cannot say that *M* always remains constant in length. It does, however, make sense to say that the two bodies are rigid *with respect to each other*. If we choose *M* as a measuring rod, we find that segments marked on *M′* remain constant in length. If we choose *M′* as a measuring rod, segments on *M* remain constant. What we have here is a concept of relative rigidity, the rigidity of one body with respect to another.

When we examine the various bodies in the world, we find many that are not rigid with respect to each other. Consider, for example, my two hands. I bring them together so that certain pairs of points on the tips of my fingers coincide. I bring them together again. The positions of my fingers have changed. The same pairs of points are no longer congruent, so I cannot say that my hands have remained rigid with respect to each other. The same is true if we compare two bodies made of wax or one body of iron and one of soft rubber. They are not rigid with respect to each other. But, just as we found that the world contains a very large class of processes that are equivalent in their periodicity, so we encounter another fortunate accidental circumstance of nature. We find, empirically, that there is one very comprehensive class of bodies that are approximately rigid relative to each other. Any two bodies of metal—iron, copper, and so on—are rigid relative to each other; so are stone bodies and even wood, if it has been well dried and is no longer green. We find that a great many solid substances are of such a kind that bodies made of those substances are rigid with respect to each other. Of course, they are not rigid if we bend them or cause them to expand by heating them, and so on. But so long as no abnormal circumstances interfere, these bodies behave in an extremely regular way as far as their lengths are concerned. When we make rough comparisons of one with another, we find them relatively rigid.

You will remember that, in our discussion of periodicity, we saw that there is no logical reason compelling us to base our measurement of time on one of the periodic processes belonging to the large class of equivalent processes. We chose such a process only because the choice resulted in a greater simplicity in our natural laws. A similar choice is involved here. There is no logical necessity for basing the measurement of length on a member of the one large class of relatively rigid bodies. We choose such bodies because it is more convenient to do so. If we chose to take a rubber or wax rod as our unit of length, we would find very few, if any, bodies in the world that were relatively rigid to our standard. Our description of nature would, therefore, become enormously complicated. We would have to say, for example, that iron bodies were constantly changing their lengths, because, each time we measured them with our flexible rubber yardstick, we obtained a different value. No scientist, of course, would want to be burdened with the complex physical laws that would have to be devised in order to describe such phenomena. On the other hand, if we choose a metal bar as a standard of length, we find that a very large number of bodies in the world are rigid when measured with it. Much greater regularity and simplicity is thus introduced into our description of the world.

This regularity derives, of course, from the nature of the actual world. We might live in a world in which iron bodies were relatively rigid to each other, and copper bodies were relatively rigid to each other, but an iron body was not relatively rigid to a copper one. There is no logical contradiction. It is a possible world. If we lived in such a world and discovered that it contained a great deal of both copper and iron, how would we choose between the two as a suitable basis for measurement? Each choice would have a disadvantage. If other metals were similarly out of step, so to speak, with each other, we would have still more difficult choices to make. Fortunately, we live in a world where this is not the case. All metals are relatively rigid to each other; therefore, we may take any one of them as our standard. When we do, we find that other metal bodies are rigid.

It is so obviously desirable to base our measurement of length on a metal rather than a rubber rod and to base our measurement of time on a pendulum rather than a pulse beat that we tend to forget that there is a conventional component in our choice of a standard. It is a component that I stressed in my doctor's thesis on space,[1] and Reichenbach later stressed in his book on space and time. The choice is conventional in the sense that there is no logical reason to prevent us from choosing the rubber rod and the pulse beat and then paying the price by developing a fantastically complex physics to deal with a world of enormous irregularity. This does not, of course, mean that the choice is arbitrary, that one choice is just as good as any other. There are strong practical grounds, the world being what it is, for preferring the steel rod and the pendulum.

Once we have chosen a standard of measurement, such as a steel rod, we are faced with another choice. We can say that the length of this particular rod is our unit, regardless of changes in its temperature, magnetism, and so on, or we can introduce correction factors depending on such changes. The first choice obviously gives the simpler rule, but if we adopt it we are again confronted by strange consequences. If the rod is heated and then used for measurement, we find that all other bodies in the world have shrunk. When the rod cools, the rest of the world expands again. We would be compelled to formulate all sorts of bizarre and complicated laws, but there would be no logical contradiction. For that reason, we can say it is a possible choice.

The second procedure is to introduce correction factors. Instead of stipulating that the segment between the two marks will always be taken as having the selected length l_0 (say, 1 or 100), we now decree that it has the normal length l_0 only when the rod is at the temperature T_0,

[1] *Der Raum. Ein Beitrag zur Wissenschaftslehre* (Jena: University of Jena, 1921); (Berlin: Verlag von Reuther & Reichard, 1922).

which we have selected as the "normal" temperature, while at any other temperature T, the length of the segment is given by the equation:

$$l = l_0 \left[1 + \beta(T-T_0)\right].$$

where β is a constant (called the "coefficient of thermal expansion") that is characteristic of the rod's substance. Similar corrections are introduced in the same way for other conditions, such as the presence of magnetic fields, that may also affect the rod's length. Physicists much prefer this more complicated procedure—the introduction of correction factors—for the same reason that they chose a metal rod instead of a rubber one—the choice leads to a vast simplification of physical laws.

CHAPTER 10

Derived Magnitudes
and the
Quantitative Language

WHEN RULES OF measurement have been given for some magnitudes, like spatial length, length of time, and mass, then, on the basis of those "primitive" magnitudes, we can introduce other magnitudes by definition. These are called "defined" or "derived" magnitudes. The value of a derived magnitude can always be determined indirectly, with the help of its definition, from the values of the primitive magnitudes involved in the definition.

In some cases, however, it is possible to construct an instrument that will measure such a magnitude directly. For example, density is commonly regarded as a derived magnitude because its measurement rests on the measurement of the primitive magnitudes length and mass. We directly measure the volume and mass of a body and then define its density as the quotient of the mass divided by the volume. It is possible, however, to measure the density of liquid directly, by means of a hydrometer. This is usually a glass float, with a long thin stem like a thermometer. The stem is marked with a scale that indicates the depth to which the instrument sinks in the liquid being tested. The liquid's approximate density is determined directly by the reading on this scale.

Thus, we find that the distinction between primitive and derived magnitudes must not be regarded as fundamental; it is rather a distinction resting on the practical procedures which physicists adopt in making measurements.

If a body is not homogeneous, we must speak of a "mean density". One is tempted to say that the density of such a body, at any given point, should be expressed as a limit of the quotient of mass divided by volume, but, because matter is discrete, the concept of limit cannot be applied here. In the cases of other derived magnitudes, the limit approach is necessary. For instance, consider a body moving along a path. During a time interval of length Δt, it travels a spatial length of Δs. We now define its "velocity", another derived magnitude, as the quotient $\Delta s / \Delta t$. If the velocity is not constant, however, we can say only that its "mean velocity" during this time interval was $\Delta s / \Delta t$. What was the velocity of the body at a certain time point during this interval? The question cannot be answered by defining velocity as a simple quotient of distance divided by time. We must introduce the concept of a limit of the quotient as the time interval approaches zero. In other words, we must make use of what in calculus is called the derivative. Instead of the simple quotient $\Delta s / \Delta t$, we have the derivative:

$$\frac{ds}{dt} = \text{limit} \frac{\Delta s}{\Delta t} \text{ for } \Delta t \to 0$$

This is called the object's "instantaneous velocity" because it expresses a velocity at a particular time point, rather than a velocity averaged over an interval. It is, of course, another example of a derived magnitude. Like the concept of density, it, too, can be measured directly by means of certain instruments; for example, the speedometer of a car provides a direct measurement of the car's instantaneous velocity.

The concept of limit is also used for defining the derived magnitude of acceleration. We have a velocity v and a change in that velocity, Δv, which occurs from one time point to another. If the interval of time is Δt and the change in velocity is Δv, the acceleration, or rate at which the velocity changes, is $\Delta v / \Delta t$. Here again, we must regard this as the "mean acceleration" during the time interval Δt. If we wish to be more precise and speak of "instantaneous acceleration" at a given time point, we must abandon the quotient of two finite values and write the following derivative:

$$\frac{dv}{dt} = \text{limit} \frac{\Delta v}{\Delta t} \text{ for } \Delta t \to 0$$

Instantaneous acceleration, therefore, is the same as the second derivative of s with respect to t:

$$a = \frac{dv}{dt} = \frac{d^2s}{dt^2}$$

At times, a physicist may say that the density of a certain point in a physical body is the derivative of its mass with respect to its volume, but this is only a rough manner of speaking. His statement cannot be taken literally, because, although space and time are (in present-day physics) continuous, the distribution of mass in a body is not—at least, not on the molecular or atomic level. For this reason, we cannot speak literally of density as a derivative; it is not a derivative in the way that this limit concept can be applied to genuinely continuous magnitudes.

There are many other derived magnitudes in physics. To introduce them, we do not have to lay down complicated rules, such as those discussed earlier for introducing primitive magnitudes. We have only to define how the derived magnitude can be calculated from the values of primitive magnitudes, which can be measured directly.

A perplexing problem sometimes arises concerning both primitive and derived magnitudes. To make it clear, imagine that we have two magnitudes, M_1 and M_2. When we examine the definition of M_1 or the rules that tell us how to measure it, we find that magnitude M_2 is involved. When we turn to the definition or rules for M_2, we find that M_1 is involved. At first, this gives the impression of circularity in the procedures, but the circle is easily avoided by applying what is called the method of successive approximation.

You will recall that in a previous chapter we considered the equation that defines the length of a measuring stick. In that equation a correction factor for thermal expansion occurs; in other words, temperature is involved in the set of rules used for measuring length. On the other hand, you will remember that in our rules for measuring temperature we referred to length, or rather, to the volume of a certain test liquid used in the thermometer; but, of course, volume is determined with the help of length. So it seems that here we have two magnitudes, length and temperature, each dependent on the other for its definition. It appears to be a vicious circle, but, in fact, it is not.

One way out is as follows. First, we introduce the concept of length without considering the correction factor for thermal expansion. This concept will not give us measurements of great precision, but it will do well enough if great precision is not demanded. For instance, if an iron rod is used for measurement, the thermal expansion, under normal conditions, is so small that measurements will still be fairly precise. This provides a first concept, L_1, of spatial length. We can now make use of this concept in the construction of a thermometer. With the aid of the iron measuring stick, we mark a scale alongside the tube containing our test liquid. Since we can construct this scale with fair precision, we also obtain a fair precision when we measure temperature on this scale. In such a way we introduce our first concept of temperature, T_1. Now we can use T_1 for establishing a refined concept of length, L_2. We do this by introducing T_1 into the rules for defining length. The refined concept of length, L_2 (corrected for the thermal expansion of the iron rod), is now available for constructing a more precise scale for our thermometer. This leads, of course, to T_2, a refined concept of temperature.

In the case of length and temperature, the procedure just described will refine both concepts to the point at which errors are extremely minute. In other cases, it may be necessary to shuttle back and forth several times before the successive refinements lead to measurements precise enough for our purposes. It must be admitted that we never reach an absolutely perfect method of measuring either concept. We can say, however, that the more we repeat this procedure—starting with two rough concepts and then refining each with the help of the other—the more precise our measurements become. By this technique of successive approximations we escape from what seems, at first, to be a vicious circle.

We will now take up a question that has been raised many times by philosophers: can measurements be applied to every aspect of nature? Is it possible that certain aspects of the world, or even certain kinds of phenomena, are in principle nonmeasurable? For instance, some philosophers may admit that everything in the physical world is measurable (although some philosophers deny even that), but they think that, in the mental world, this is not the case. Some even go so far as to contend that everything that is mental is not measurable.

A philosopher who takes this point of view might argue as follows: "The intensity of a feeling or of a bodily pain or the degree of intensity with which I remember a past event is in principle not measurable. I may feel that my memory of one event is more intense than my memory of another, but it is not possible for me to say that one is intense to the degree of 17 and the other to the degree of 12.5. Measurement of intensity of memory is, therefore, in principle impossible."

In reply to this point of view, let us first consider the physical magnitude of weight. You pick up a stone. It is heavy. You compare it with another stone, a much lighter one. If you examine both stones, you will not come upon any numbers or find any discrete units that can be counted. The phenomenon itself contains nothing numerical—only your private sensations of weight. As we have seen in a previous chapter, however, we introduce the numerical concept of weight by setting up a procedure for measuring it. It is *we* who assign numbers to nature. The phenomena themselves exhibit only qualities that we observe. Everything numerical, except for the cardinal numbers that can be correlated with discrete objects, is brought in by ourselves when we devise procedures for measurement.

The answer to our original philosophic question should, I think, be put this way. If, in any field of phenomena, you find sufficient order that you can make comparisons and say that, in some respect, one thing is above another thing, and that one is above something else, then there is in principle the possibility of measurement. It is now up to you to devise rules by which numbers can be assigned to the phenomena in a useful way. As we have seen, the first step is to find comparative rules; then, if possible, to find quantitative ones. When we assign numbers to phenomena, there is no point in asking whether they are the "right" numbers. We simply devise rules that specify how the numbers are to be assigned. From this point of view, nothing is in principle unmeasurable.

Even in psychology we do, in fact, measure. Measurements for sensation were introduced in the nineteenth century; perhaps the reader recalls the Weber-Fechner law, in what was then called the field of psycho-physics. The sensation to be measured was first correlated with something physical; then rules were laid down to determine the degree of intensity of the sensation. For example, measurements were made of the feeling of the pressure on the skin of various weights, or the sensation of the pitch of a sound, or the intensity of a sound, and so on. One approach to measuring pitch—we are here speaking about the sensation, not the frequency of the sound wave—is to construct a scale based on a unit that is the smallest difference in pitch that one can detect. S. S. Stevens, at one time, proposed another procedure, based on a subject's identification of a pitch, which he felt to be exactly midway between two other pitches. So, in various ways, we have been able to devise measuring scales for certain psychological magnitudes. It is certainly not the case, therefore, that there is in principle a fundamental impossibility of applying the quantitative method to psychological phenomena.

At this point, we should comment on a limitation of the procedure of measurement. There is not the slightest doubt, of course, that measurement is one of the basic procedures of science, but, at the same time, we must be careful not to overestimate its scope. The specification of a measuring procedure does not always give us the whole meaning of a concept. The more we study a developed science, especially such a richly developed one as physics, the more we become aware of the fact that the total meaning of a concept cannot be given by one measuring procedure. This is true of even the simplest concepts.

As an example, consider spatial length. The procedure of measuring length with a rigid rod can be applied only within a certain intermediate range of values that are not too large and not too small. It can be applied to as small a length as, perhaps, a millimeter or a fraction of a millimeter, but not to a thousandth of a millimeter. Extremely small lengths cannot be measured in this fashion. Nor can we apply a measuring rod to the distance from the earth to the moon. Even the distance from the United States to England cannot be measured by such a procedure without first building a solid bridge from here to England. Of course, we continue to speak of a spatial distance between this country and England, meaning a distance that *could* be measured with a measuring rod if the earth's surface between the two countries were solid. But the surface is not solid, so, even here, we must devise other procedures for measuring length.

One such procedure is as follows. By means of a measuring rod we establish a certain distance on the earth, for example, between the points A and B (see Figure 10–1). With this line AB as a basis, we can determine the distance from B to a remote point C, without using a measuring rod. By means of surveying instruments, we measure the two angles, α and β. Theorems of physical geometry enable us to calculate the length of line a, which is the distance between B and C. Knowing this distance, and measuring angles δ and γ, we can calculate the distance from B to an even more remote point D. Thus, by the process called "triangulation", we can measure a large net of distances and in this way construct a map of a large region.

Astronomers also use triangulation for measuring distances from the earth to relatively near stars within our galaxy. Of course, distances on the earth are much too short to be used as base lines, so astronomers use the distance from one point of the earth's orbit to the opposite point.

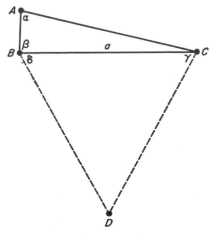

Figure 10–1.

This method is not accurate enough for stars at very great distances within our galaxy or for measuring distances to other galaxies, but, for such enormous distances, other methods can be used. For example, the intrinsic brightness of a star can be determined from its spectrum; by comparing this with the star's brightness as observed from the earth, its distance can be estimated. There are many ways to measure distances that cannot be measured by the direct application of a measuring rod. We observe certain magnitudes and then, on the basis of laws connecting these magnitudes with other magnitudes, we arrive at indirect estimates of distances.

At this point, an important question arises. If there are a dozen different ways to measure a certain physical magnitude, such as length, then, instead of a single concept of length, should we not speak of a dozen different concepts? This was the opinion expressed by the physicist and philosopher of science P. W. Bridgman in his now-classic work, *The Logic of Modern Physics* (Macmillan, 1927). Bridgman stressed the view that every quantitative concept must be defined by the rules involved in the procedure for measuring it. This is sometimes called an "operational definition" of a concept. But, if we have many different operational definitions of length, we should not, according to Bridgman, speak of *the* concept of length. If we do, we must abandon the notion that concepts are defined by explicit measuring procedures.

My view on this question is as follows. I think it is best to regard the concepts of physics as theoretical concepts in the process of being specified in stronger and stronger ways, not as concepts completely defined by operational rules. In everyday life, we make various observations of nature. We describe those observations in qualitative terms, such as "long", "short", "hot", "cold", and in comparative terms, such as "longer", "shorter", "hotter", "colder". This observation language is connected with the theoretical language of physics by certain operational rules. In the theoretical language, we introduce quantitative concepts such as length and mass, but we must not think of such concepts as explicitly defined. Rather, the operational rules, together with *all* the postulates of theoretical physics, serve to give partial definitions, or rather, partial interpretations of the quantitative concepts.

We know that these partial interpretations are not final, complete definitions, because physics is constantly strengthening them by new laws and new operational rules. No end to this process is in sight—physics is far from having developed a complete set of procedures—so we must admit that we have only partial, incomplete interpretations of all the theoretical terms. Many physicists include such terms as "length" in the observation vocabulary because they can be measured by simple, direct procedures. I prefer not to classify them this way. It is true that, in everyday language, when we say, "The length of this edge of the table is thirty inches", we are using "length" in a sense that can be completely defined by the simple measuring-rod procedure. But that is only a small part of the total meaning of the concept of length. It is a meaning that applies only to a certain intermediate range of values to which the measuring-rod technique can be applied. It cannot be applied to the distance between two galaxies or between two molecules. Yet clearly, in these three cases, we have in mind the same concept. Instead of saying that we have many concepts of length, each defined by a different operational procedure, I prefer to say that we have one concept of length, partially defined by the entire system of physics, including the rules for all the operational procedures used for the measurement of length.

The same is true for the concept of mass. If we restrict its meaning to a definition referring to a balance scale, we can apply the term to only a small intermediate range of values. We cannot speak of the mass of the moon or of a molecule or even the mass of a mountain or of a house. We should have to distinguish between a number of different magnitudes, each with its own operational definition. In cases in which two different methods for measuring mass could be applied to the same object, we would have to say that, in those cases, the two magnitudes happened to have the same value. All this would lead, in my opinion, to an unduly complicated way of speaking. It seems best to adopt the language form used by most physicists and regard length, mass, and so on as theoretical concepts rather than observational concepts explicitly defined by certain procedures of measurement.

This approach is no more than a matter of preference in the choice of an efficient language. There is not just one way to construct a language of science. There are hundreds of different ways. I can say only that, in my view, this approach to the quantitative magnitudes has many advantages. I have not always held this view. At one time, in agreement with many physicists, I regarded concepts such as length and mass as "observables"—terms in the observation language. But, more and more, I am inclined to enlarge the scope of the theoretical language and include in it such terms. Later we shall discuss theoretical terms in more detail. Now I only want to point out that, in my view, the various procedures of measurement should not be thought of as defining magnitudes in any final sense. They are merely special cases of what I call "correspondence rules". They serve to connect the terms of the observation language with the terms of the theoretical language.

CHAPTER 11

Merits of the
Quantitative Method

QUANTITATIVE CONCEPTS are not given by nature; they arise from our practice of applying numbers to natural phenomena. What are the advantages of doing this? If the quantitative magnitudes were supplied by nature, we would no more ask this question than we would ask: what are the advantages of colors? Nature might not have colors, but it is pleasant to find them in the world. They are simply there, a part of nature. We cannot do anything about it. The situation is not the same with respect to the quantitative concepts. They are part of our language, not part of nature. It is *we* who introduce them; therefore, it is legitimate to ask *why* we introduce them. Why do we go to all the trouble of devising complicated rules and postulates in order to have magnitudes that be measured on numerical scales?

We all know the answer. It has been said many times that the great progress of science, especially in the last few centuries, could not have occurred without the use of the quantitative method. (It was first introduced in a precise way by Galileo. Others had used the method earlier, of course, but he was the first to give explicit rules.) Wherever it is possible, physics tries to introduce quantitative concepts. In the last decades, other fields of science have followed the same path. We have no doubt that this is advantageous, but it is good to know in greater detail exactly where the advantages lie.

First of all—though this is only a minor advantage—there is an increase in the efficiency of our vocabulary. Before a quantitative concept is introduced, we have to use dozens of different qualitative terms or adjectives in order to describe the various possible states of an object with respect to that magnitude. Without the concept of temperature, for example, we have to speak of something as "very hot", "hot", "warm", "lukewarm", "cool", "cold", "very cold", and so on. These are all what we have called classificatory concepts. If we had a few hundred such terms, perhaps it would not be necessary, for many everyday purposes, to introduce the quantitative concept of temperature. Instead of saying, "It is 95 degrees today", we would have a nice adjective that meant just this temperature, and for 100 degrees we would have another adjective, and so on.

What would be wrong with this? For one thing, it would be exceedingly hard on our memory. We would not only have to know a great number of different adjectives, but we would also have to memorize their order, so we would know immediately whether a certain term was higher or lower on the scale than another. But, if we introduce the single concept of temperature, which correlates the states of a body with numbers, we have only one term to memorize. The order of magnitude is immediately supplied by the order of the numbers. It is true, of course, that we must have previously memorized the numbers, but once we have done so, we can apply those numbers to any quantitative magnitude. Otherwise, we should have to memorize a different set of adjectives for every magnitude, and, in each case, we should also have to memorize their specific order. These are two minor advantages of the quantitative method.

The major advantage, as we have seen in previous chapters, is that quantitative concepts permit us to formulate quantitative laws. Such laws are enormously more powerful, both as ways to explain phenomena and as means for predicting new phenomena. Even with an enriched qualitative language, in which our memory is burdened with hundreds of qualifying adjectives, we would have great difficulty expressing even the simplest laws.

Suppose, for instance, that we have an experimental situation in which we observe that a certain magnitude P is dependent on a certain other magnitude *M.*We plot this relation as the curve shown in Figure 11–1. On the horizontal line of this graph, magnitude *M* assumes the

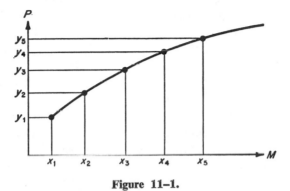

Figure 11–1.

values $x_1, x_2, \ldots$ For those values of *M*, magnitude *P* takes the values $y_1, y_2, \ldots$ After plotting on the graph the points that pair these values, we try to fit a smooth curve through these points. Perhaps they fit a straight line; in that case, we say that *P* is a linear function of *M* We express this as $P = aM + b$, where *a* and *b* are parameters that remain constant in the given situation. If the points fit a second degree curve, we have a quadratic function. Perhaps *P* is the logarithm of *M*, or it may be a more complicated function that must be expressed in terms of several simple functions. After we have decided on the most likely function, we test, by repeated observations, whether we have found a function that represents a universal law connecting the two magnitudes.

What would happen in this situation if we did not have a quantitative language? Assume that we have a qualitative language far richer than present-day English. We do not have such words as "temperature" in our language, but we do have, for every quality, some fifty adjectives, all neatly ordered. Our first observation would not be $M = x_1$. Instead, we would say that the object we are observing is ———, using here one of the fifty adjectives that refer to *M*. And, instead of $P = y_1$, we would have another sentence in which we employ one of the fifty adjectives that have reference to the quality *P*. Strictly speaking, the two adjectives would not correspond to points on the axes of our graph—we could not possibly have enough adjectives to correspond to *all* the points on a line—but rather to intervals along each line. One adjective, for example, would refer to an interval that contained x_1. The fifty intervals alone the axis for M, corresponding to our fifty adjectives for M, would have fuzzy boundaries; they might even overlap to some extent. We could not, in this language, express a simple law of, say, the form $P = a + bM + cM^2$. We would have to specify exactly how each of our fifty adjectives for M is to be paired with one of the fifty adjectives for P.

To be more specific, suppose that M refers to heat qualities, and P refers to colors. A law connecting these two qualities would consist of a set of fifty conditional sentences of the form, "If the object is very, very, very hot (of course, we would have one adjective to express this), then it is bright red." Actually, we do have in English quite a large number of adjectives for colors, but that is almost the only field of qualities for which we have many adjectives. In reference to most of the magnitudes in physics, there is a great paucity of adjectives in the qualitative language. A law expressed in a quantitative language is thus much shorter and simpler than the cumbersome expressions that would be required if we tried to express the same law in qualitative terms. Instead of one simple, compact equation, we would have dozens of "if-then" sentences, each pairing a predicate of one class with a predicate of another.

The most important advantage of the quantitative law, however, is not its brevity, but rather the use that can be made of it. Once we have the law in numerical form, we can employ that powerful part of deductive logic we call mathematics and, in that way, make predictions. Of course, in the qualitative language, deductive logic could also be used for making predictions. We could deduce from the premiss, "This body will be very, very, very hot", the prediction, "This body will be bright red." But the procedure would be cumbersome compared to the powerful, efficient methods of deduction that are part of mathematics. This is the greatest advantage of the quantitative method. It permits us to express laws in a form using mathematical functions by which predictions can be made in the most efficient and precise way.

These advantages are so great that no one today would think of proposing that physicists abandon the quantitative language and return to a prescientific qualitative language. In earlier days of science, however, when Galileo was calculating the speeds with which balls rolled down inclined planes and the periods of a pendulum, there were many who probably said: "What good will come of all this? How will it help us in everyday life? I shall never be concerned with what happens to small spherical bodies when they roll down a track. It is true that sometimes, when I am shelling peas, they run down an inclined table. But what is the value of calculating their *exact* acceleration? What practical use could such knowledge have?"

Today, no one speaks this way because we are all using dozens of complicated instruments—a car, a refrigerator, a television set—which we know would not have been possible if physics had not developed as a quantitative science. I have a friend who once took the philosophical attitude that the development of quantitative science was regrettable because it led to a mechanization of life. My reply was that, if he wished to be consistent in this attitude, he should never use an airplane or a car or a telephone. To abandon quantitative science would mean the abandonment of all those conveniences that are products of modern technology. Not many people, I believe, would wish to do that.

At this point, we face a related, though somewhat different, criticism of the quantitative method. Does it really help us to *understand* nature? Of course, we can describe phenomena in mathematical terms, make predictions, invent complicated machines; but are there not better ways to obtain true insights into nature's secrets? Such a criticism of the quantitative method as inferior to a more direct, intuitive approach to nature was made by the greatest of German poets, Goethe. The reader probably knows him only as a writer of drama and poetry, but actually he was much interested in certain parts of science, particularly in biology and the theory of colors. He wrote a large book on the theory of colors. At times, he believed that that book was more important than all his poetic works put together.

A portion of Goethe's book deals with the psychological effects of colors. It is systematically presented and really quite interesting. Goethe was sensitive in observing his experiences and, for that reason, was well qualified to discuss how our moods are influenced by the colors surrounding us. Every interior decorator, of course, knows these effects. A lot of yellow and red in a room is stimulating. Greens and blues have a calming effect. When we select colors for our bedrooms and living rooms, we keep those psychological effects in mind. Goethe's book also takes up the physical theory of color; there is a historical section in which he discusses previous theories, especially Newton's theory. He was dissatisfied in principle with Newton's entire approach. The phenomena of light in all its aspects, Goethe contended, especially in its color aspects, should be observed only under the most natural conditions. His work in biology had led him to conclude that, if you want to find out the real character of an oak tree or a fox, you must observe the

tree and the fox in their natural habitats. Goethe transferred this notion
to physics. One observes a thunderstorm best by going out during a
thunderstorm and looking at the sky. So also with light and colors. One
must see them as they occur in nature—the way sunlight breaks through
a cloud, how the colors of the sky alter when the sun is setting. By doing this, Goethe found some regularities. But, when he read in Newton's
famous book, *Opticks,* the assertion that white light from the sun is
actually a compound of all the spectral colors, Goethe was much incensed.

Why was he incensed? Because Newton did not make his observations of light under natural conditions. Instead, he made his famous experiment indoors, with a prism. He darkened his laboratory and cut a
tiny slit in the window shutter (see Figure 11–2), a slit allowing only a

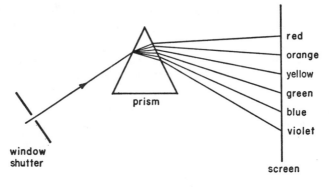

Figure 11–2.

narrow beam of sunlight to enter the dark room. When this ray of light
passed through a prism, Newton observed that it cast on the screen a
pattern of different colors, from red to violet. He called this pattern a
spectrum. By measuring the angles of refraction at the prism, he concluded that those angles were different for different colors, smallest for
red, largest for violet. This led him to the assumption that the prism does
not produce the colors; it merely separates colors contained in the original beam of sunlight. He confirmed this assumption by other experiments.

Goethe raised several objections to Newton's general approach to
physics, as illustrated by this experiment. First, he said, in trying to
understand nature, we should rely more on the immediate impression

our senses receive than on theoretical analysis. Since white light appears to our eye as perfectly simple and colorless, we should accept it as such and not represent it as composed of different colors. It also seemed wrong to Goethe to observe a natural phenomenon, such as sunlight, under artificial, experimental conditions. If you want to understand sunlight, you must not darken your room and then squeeze the beam of light through a narrow slit. You should go out under the open sky and contemplate all the striking color phenomena as they appear in their natural setting. Finally, he was sceptical about the usefulness of the quantitative method. To take exact measurements of angles, distances, speeds, weights, and so on and then make mathematical calculations based on the results of these measurements, might be useful, he conceded, for technical purposes. But he had serious doubts as to whether this was the best approach if we wish to gain real insight into nature's ways.

Today, of course, we know that, in the controversy between Newton's analytical, experimental, quantitative method and Goethe's direct, qualitative, phenomenological approach, the former has not only won out in physics, but today is gaining more and more ground in other fields of science as well, including the social sciences. It is now obvious, especially in physics, that the great advances of the last centuries would not have been possible without the use of quantitative methods.

On the other hand, we should not overlook the great value that an intuitive approach like Goethe's may have for the discovery of new facts and the development of new theories, especially in relatively new fields of knowledge. Goethe's way of artistic imagination, combined with careful observation, enabled him to discover important new facts in the comparative morphology of plant and animal organisms. Some of these discoveries were later recognized as steps in the direction of Darwin's theory of evolution. (This was explained by the great German physicist and physiologist Hermann von Helmholtz in a lecture in 1853 on Goethe's scientific studies. Helmholtz praised highly Goethe's work in biology, but he criticized his theory of colors. In an 1875 postscript to the lecture, he pointed out that some of Goethe's hypotheses had, in the meantime, been confirmed by Darwin's theory.)[1]

---

[1] Goethe's *Die Farbenlehre* ("Theory of Colors") was a massive three-part work published in Germany in 1810. An English translation of Part I, by Charles Eastlake, was issued in London in 1840. Helmholtz's lecture, "On Goethe's Scientific Researches," first appeared in English in his *Popular Lectures on Scientific Subjects* (London: Longmans, Green, 1881), and was reprinted in his *Popular Scientific Lectures* (New York: Dover, 1962). For similar criticism of

It may be of interest to mention that, near the middle of the last century, the philosopher Arthur Schopenhauer wrote a little treatise on vision and colors (*Über das Sehn und die Farben*) in which he took the position that Goethe was entirely right and Newton entirely wrong in their historic controversy. Schopenhauer condemned not only the application of mathematics to science, but also the technique of mathematical proofs. He called them "mouse-trap proofs", citing as an example the proof for the familiar theorem of Pythagoras. This proof, he said, is correct; no one can contradict you and say it is wrong. But it is an entirely artificial way of reasoning. Every step is convincing, of course, yet at the conclusion of the proof you have the feeling that you have been caught in a mouse trap. The mathematician has compelled you to admit the truth of the theorem, but you have gained no real understanding. It is as if you had been led through a maze. You suddenly walk out of the maze and say to yourself: "Yes, I am here, but I really do not know how I got here." There is something to be said for this point of view in the teaching of mathematics. We should give more attention to the intuitive understanding of what we are doing at each step along the way in a proof, and why we are taking those steps. But all this is by the way.

To give a clear answer to the question of whether, as some philosophers believe, we lose something when we describe the world with numbers, we must clearly distinguish between two language situations: a language that actually does leave out certain qualities of the objects it describes and a language that *seems* to leave out certain qualities but actually does not. I am convinced that much of the confusion in the thinking of these philosophers results from a failure to make this distinction.

"Language" is used here in an unusually wide sense. It refers to any method by which information about the world is communicated—words, pictures, diagrams, and so on. Let us consider a language that leaves out certain aspects of the objects it describes. You see in a magazine a black and white photograph of Manhattan. Perhaps the caption reads: "New York's skyline, seen from the west." This picture com-

---

Goethe, see "Goethe's 'Farbenlehre,' " an address by John Tyndall, in his *New Fragments* (New York: Appleton, 1892), and Werner Heisenberg's 1941 lecture, "The Teachings of Goethe and Newton on Colour in the Light of Modern Physics," in *Philosophic Problems of Nuclear Science* (London: Faber & Faber, 1952).

municates, in the language of black and white photography, information about New York. You learn something of the sizes and shapes of
the buildings. The picture is similar to the immediate visual impression
you would have if you stood where the camera stood and looked toward
New York. That, of course, is why you immediately understand the
picture. It is not language in the ordinary sense of the word; it is language in the more general sense that it conveys information.

Yet the photograph lacks a great deal. It does not have the dimension of depth. It tells you nothing about the colors of the buildings. This
does not mean that you cannot make correct inferences about depth and
color. If you see a black and white photograph of a cherry, you assume
that the cherry is probably red. But this is only an inference. The picture itself does not communicate the color of the cherry.

Now let us turn to the situation in which qualities appear to be left
out of a language when actually they are not. Consider a sheet of music.
When you first saw musical notation, perhaps as a child, you may have
asked: "What are these strange things here? There are five lines that
stretch across the page, and they are covered with black spots. Some of
the spots have tails."

You were told: "This is music. This is a very beautiful melody."

You protest: "But I can't hear any music."

It is certainly true that this notation does not convey a melody in
the same way that, say, a phonograph record does. There is nothing to
hear. In another sense, however, the notation *does* convey the pitch and
duration of each tone. It is not conveyed in a way that is meaningful to
a child. Even to an adult the melody may not be immediately apparent
until he has played it on a piano or asked someone else to play it for
him; yet there is no doubt that the tones of the melody are implicit in
the notation. Of course, a translation key is needed. There must be rules
about how to transform this notation into sounds. But, if the rules are
known, we can say that the qualities of the tones—their pitch, duration,
even intensity changes—are given in the notation. A trained musician
may even be able to scan the notes and "hear" the melody immediately
in his mind. We obviously have here a language situation clearly different from that of the black and white photograph. The photograph actually left out colors. The musical notation seems to leave out tones, but
actually does not.

In the case of ordinary language, we are so accustomed to the
words that we often forget that they are not natural signs. If you hear

the word "blue", you immediately imagine the color blue. As children, we form the impression that the color words of our language actually do convey color. On the other hand, if we read a statement by a physicist that there is a certain electromagnetic oscillation of a certain intensity and frequency, we do not immediately imagine the color it describes. If you know the translation key, however, you can determine the color just as accurately, perhaps even more accurately, than if you heard the color word. If you have done work yourself with a spectroscope, you may know by heart which colors correspond to which frequencies. In that case, the physicist's statement may tell you immediately that he is speaking about a color that is blue-green.

The translation key may be laid down in many different ways. For example, the frequency scale of the visible spectrum can be recorded on a chart, and the English color word most closely corresponding to each frequency is written after it. Or the chart may have, instead of the color words, small squares containing the actual colors. In either case, when you hear the physicist's quantitative statement, you can infer, with the help of the key, exactly what color he is describing. The quality, in this case the color, is not at all lost by his method of communication. The situation here is analogous to that of musical notation; there is a key for determining those qualities that seem, at first, to be omitted from the notation. It is not analogous to the black and white photograph, in which certain qualities actually are left out.

The advantages of the quantitative language are so obvious that one is led to wonder why so many philosophers have criticized its use in science. In Chapter 12, we will discuss some of the reasons for this curious attitude.

CHAPTER 12

The Magic View
of Language

I HAVE THE IMPRESSION that one reason why some philosophers object to the emphasis science places on quantitative language is that our psychological relation to the words of a prescientific language—words we learned when we were children—is quite different from our psychological relation to those complicated notations we later come upon in the language of physics. It is understandable how children can believe that certain words actually do carry, so-to-speak, the qualities to which they refer. I do not want to be unfair to certain philosophers, but I suspect that those philosophers sometimes make the same mistake, in their reactions to scientific words and symbols, that children always make.

In the well-known book by C. K. Ogden and I. A. Richards, *The Meaning of Meaning*,[1] there are excellent examples—some are quite amusing—of what the authors call "word magic". Many people hold a magical view of language, the view that there is a mysterious natural

[1] C. K. Ogden and I. A. Richards, *The Meaning of Meaning* (London: Kegan Paul, Trench, Trubner, 1923); (8th rev. ed.; New York: Harcourt, Brace, 1946); (New York: Harvest Books, 1960).

115

connection of some sort between certain words (only, of course, the words with which they are familiar!) and their meanings. The truth is, that it is only by historical accident, in the evolution of our culture, that the word "blue" has come to mean a certain color. In Germany that color is called "blau". Other sounds are associated with it in other languages. For children, it is natural to think that the *one* word "blue", to which they are accustomed in their mother tongue, is the natural word, that other words for blue are entirely wrong or certainly strange. As they grow older, they may become more tolerant and say: "Other people may use the world 'blau', but they use it for a thing that is *actually* blue." A small boy thinks that a house is a house, and a rose is a rose, and that is all there is to it. Then he learns that the strange people in France call a house a "maison". Why do they say "maison" when they mean a house? Since it *is* a house, why don't they call it a house? He will be told that it is the custom in France to say "maison". Frenchmen have been saying it for hundreds of years; he should not blame them for it or think them stupid. The boy finally accepts this. The strange people have strange habits. Let them use the word "maisons" for those things that are actually houses. To break away from this tolerant attitude and acquire the insight that there is no essential connection whatever between a word and what we mean by it seems to be as difficult for many adults as it is for children. Of course, they never say openly that the English word is the right word, that words in other languages are wrong, but the magical view of their childhood remains implicit in their thinking and often in their remarks.

Ogden and Richards quote an English proverb, "The Divine is rightly so called." This apparently means that the Divine is really divine; therefore, he is rightly so called. Although one may have the feeling that something is rightly so called, the proverb does not, in fact, say anything. It is obviously vacuous. Nevertheless, people evidently repeat it with strong emotion, actually thinking that it expresses some sort of deep insight into the nature of the Divine.

A slightly more sophisticated example of the magic view of language is contained in a book by Kurt Riezler, *Physics and Reality: Lectures of Aristotle on Modern Physics at an International Congress of Science, 679 Olympiad, Cambridge, 1940* A.D.[2] The author imagines Aristotle coming back to earth in our time and presenting his point of

---

[2] Kurt Riezler's book was published in 1940 by Yale University Press, New Haven, who granted permission to quote from the book.

view—which is Riezler's point of view also and, I think, only Riezler's —in regard to modern science.

Aristotle begins by highly praising modern science. He is full of admiration for its great achievements. Then he adds that, to be honest about it, he must also make a few critical remarks. It is these remarks that interest us here. On page 70 of Riezler's book, Aristotle says to the assembled physicists:

> The day is cold to a Negro and hot to an Eskimo. You settle the dispute by reading 50° on your thermometer.

What Riezler wants to say here is that, in the qualitative language of daily life, we have no agreement about words like "hot" and "cold". If an Eskimo from Greenland arrives at a spot where the temperature is 50 degrees, he will say: "This is a rather hot day." A Negro from Africa, at the same spot, will say: "This is a cold day." The two men do not agree on the meanings of "hot" and "cold". Riezler imagines a physicist saying to them: "Let's forget about those words and speak instead in terms of temperature; then we can come to an agreement. We will agree that the temperature today is 50 degrees."

The quotation continues:

> You are proud of having found the objective truth, by eliminating . . .

I ask the reader to guess what Riezler thinks the physicists have eliminated. We might expect the sentence to continue, ". . . by eliminating the words 'hot' and 'cold' ". The physicist does not, of course, eliminate these words from anything but the quantitative language of physics. He will still want them in the qualitative language of everyday life. Indeed, the qualitative language is essential, even to the physicist, in order to describe what he sees. But Riezler does not go on to say what we expect. His statement continues:

> . . . by eliminating both the Negro and the Eskimo.

When I first read this I thought that he was just saying it a little differently and that he means that the physicist eliminates the ways of speaking of the Negro and the Eskimo. But this is not the case. Riezler means something much deeper. Later on he makes it quite clear that, in his view, modern science has eliminated man, has forgotten and neglected the most important of all topics of human knowledge—man himself.

You are proud of having found the objective truth, by eliminating both the Negro and the Eskimo. I grant the importance of what you have achieved. Granted, also, that you could not build your wonderful machines without eliminating the Negro and Eskimo. What about reality and truth? You identify truth with certitude. But obviously, truth is concerned with Being or, if you prefer, with something called "reality". Truth can have a high degree of certitude, as truth in mathematics surely has, and nevertheless a low degree of 'reality'. What about your 50°? Since it is true for both the Negro and Eskimo you call it objective reality. This reality of yours seems to me to be extremely poor and thin. It is a relation connecting a property called temperature with the expansion of your mercury. This reality does not depend on the Negro or the Eskimo. It is related to neither but to the anonymous observer.

Somewhat later he writes:

Of course, you are quite aware that heat and cold relate 50° to the Negro or Eskimo.

I am not quite sure what he means to say there. Perhaps he means that, if the Negro and the Eskimo are to understand what is meant by "50°", it must be explained to them in terms of "hot" and "cold".

You say that the system under observation needs to be enlarged to include the physical happenings within the Negro or Eskimo.

This is meant to be the physicist's answer to the charge: "Do you not omit the sensations of heat and cold which the Eskimo and the Negro respectively feel?" Riezler seems to think that the physicist would reply somewhat like this: "No, we don't omit the sensations. We describe also the Negro himself, and the Eskimo, as organisms. We analyze them as physical systems, physiological and physical. We find out what happens inside them, and in this way we can explain why they experience different sensations which lead them to describe the same day as 'hot' and 'cold'." The passage continues:

That confronts you with two systems in which the gradient of temperature is reversed—cold in the one and warm in the other system. This cold and warm, however, is not yet cold and warm. The Negro and the Eskimo are represented in your systems by a compound of physical or chemical happenings; they are no longer beings in themselves, they are what they are relative to the anonymous observer, a compound of happenings described by relations between measurable quantities. I feel that the Negro and Eskimo are represented in your description rather meagerly. You place the responsibility upon the enormous complications involved in such a system.

Riezler refers here to the human system; the total organism which of course is enormously complicated when you try to analyze it physically. He continues:

> No, gentlemen, you coordinate symbols but you never describe cold as cold and warm as warm.

Here, it comes out at last—at least a little suspicion of the magic of words! The physicist coordinates artificial symbols that really do not carry anything like the qualities. This is unfortunate, because the physicist is unable to describe cold as "cold". Calling it "cold" would convey to us the actual sensation. We would all shiver, just imagining how cold it was. Or, saying "Yesterday it was terribly hot" would give us the actual feeling of heat. This is my interpretation of what Riezler is saying. If the reader wishes to make a more benevolent interpretation, he is free to do so.

Later (on p. 72), there is another interesting declaration by Riezler's Aristotle:

> Let me return to my point. Reality is the reality of substances. You do not know the substances behind the happenings your thermometer represents in indicating 50°. But you know what the Negro and Eskimo are like. . . .

Riezler means that you know what the Negro and Eskimo are like because they are humans. You are a human, so you share with them common feelings.

> . . . ask them, ask yourselves, ask your pain and your joy, your acting and being acted on. There you know what reality means. There things are concrete. There you know that they *are*.

The *real* reality, he feels, can be reached only when we talk about pain and joy, hot and cold. As soon as we go over to the symbols of physics, temperature, and the like, the reality thins out. This is Riezler's judgment. I am convinced that it is not Aristotle's. Aristotle was one of the greatest men in the history of thinking; in his time he had supreme respect for science. He himself made empirical observations and experiments. If he could have observed the development of science from his day to ours, I am sure that he would be enthusiastically in favor of the scientific way of thinking and speaking. Indeed, he would probably be one of today's leading scientists. I think Riezler does considerable injustice to Aristotle in attributing these opinions to him.

It is possible, I suppose, that Riezler meant to say only that science should not concentrate so exclusively on quantitative concepts that it neglects all those aspects of nature that do not fit so neatly into formulas with mathematical symbols. If this is all that he meant, then of course we would agree with him. For example, in the field of aesthetics, there has not been much progress in the development of quantitative concepts. But it is always difficult to say in advance where it will be useful to introduce numerical measurement. We must leave this to the workers in the field. If they see a way of doing it usefully, they will introduce it. We should not discourage such efforts before they have been made. Of course, if language is used for aesthetic purposes—not as a scientific investigation of aesthetics, but for giving aesthetic pleasure—then there is no question about the unsuitability of the quantitative language. If we want to express our feelings, in a letter to a friend or in a lyric poem, naturally we choose a qualitative language. We need words so familiar to us that they immediately call up a variety of meanings and associations.

It is also true that sometimes a scientist neglects important aspects of even the phenomena on which he is working. This, however, is often just a matter of division of labor. One biologist does his work entirely in the laboratory. He studies cells under a microscope, makes chemical analyses, and so on. Another biologist goes out into nature, observes how plants grow, under what conditions birds build nests, and so on. The two men have different interests, but the knowledge they acquire in their separate ways is all part of science. Neither should suppose that the other does useless work. If Riezler's intention is merely to warn us that science should be careful not to leave out certain things, one can go along with him. But if he meant to say, as he seems to say, that the quantitative language of science actually omits certain qualities, then I think he is wrong.

Let me quote from a review of Riezler's book by Ernest Nagel.[3] "The theories of physics are not surrogates for the sun and the stars and the many-sided activities of concrete things. But why should anyone reasonably expect to be warmed by discourse?"

You see, Nagel interprets Riezler in an even less charitable way than I have tried to do. He may be right. I am not quite sure. Nagel understands Riezler as criticizing the language of the physicist for not directly conveying, in the stronger sense, qualities such as the colors that

[3] *Journal of Philosophy,* 37 (1940), 438–439.

are actually contained in a colored picture. In the same way, we could convey information about smells by spraying perfume—bringing in actual odors, rather than just naming them. Perhaps Riezler meant— Nagel understands him so—that language should convey qualities in this strong sense, that it should actually bring the qualities to us. He seems to think that a word like "cold" somehow carries with it the actual quality of coldness. Such a point of view is certainly an example of the magic view of language.

Part **III**

THE STRUCTURE
OF SPACE

CHAPTER 13

Euclid's Parallel
Postulate

THE NATURE OF geometry in physics is a topic of great importance in the philosophy of science—one, by the way, in which I have a special interest. I wrote my doctoral thesis on this subject, and, although I have published little on it since, it is a topic about which I have continued to think a great deal.

Why is it so important? First of all, it leads into an analysis of the space-time system, the basic structure of modern physics. Moreover, mathematical geometry and physical geometry are excellent paradigms of two fundamentally different ways of gaining knowledge: the aprioristic and the empirical. If we clearly understand the distinction between these two geometries, we shall obtain valuable insights into important methodological problems in the theory of knowledge.

Let us first consider the nature of mathematical geometry. We know, of course, that geometry was one of the very earliest mathematical systems to be developed. We know little about its origins. The amazing thing is that it was already so well systematized at the time of Euclid. The axiomatic character of Euclid's geometry—the deriving of theorems from fundamental axioms and postulates—was itself a remarkably sophisticated contribution, one that still plays a basic role in the most modern ways of putting mathematical systems into exact form. It is astonishing that this procedure was already being followed in Euclid's time.

One of Euclid's axioms, the axiom of the parallel, gave a great deal of trouble to mathematicians for many centuries. We can state this axiom as follows: For any plane on which there is a straight line L and a point P that is not on L, there is one and only one straight line L', on the plane, that passes through P and is parallel to L. (Two lines on a plane are defined as parallel if they have no point in common.)

This axiom seemed so obvious that, until the beginning of the last century, no one doubted its truth. The debate that centered about it was not over its truth, but over whether it was necessary as an *axiom*. It seemed to be less simple than the other axioms of Euclid. Many mathematicians believed that it might be a *theorem* that could be derived from the other axioms.

Numerous attempts were made to derive the parallel axiom from other axioms, and some mathematicians even claimed that they had succeeded. We know today that they were mistaken. It was not easy at the time to see the flaw in each of these supposed derivations because they were usually based—as they still are often based in high-school geometry textbooks—on an appeal to our intuitions. We draw a diagram. Admittedly, the diagram is inexact. There are no perfect lines— the lines we draw have a thickness because of the chalk on the blackboard or the ink on the paper—but the diagram aids our imagination. It helps us "see" the truth of what we wish to prove. The philosophy of this intuitive approach was best systematized by Immanuel Kant. It is not our sense impression of the physical diagram, but rather our inner intuition of geometrical configurations, that cannot be mistaken. Kant was quite clear about this. One can never be certain that two line segments on the blackboard are equal or that a chalk line that is supposed to be a circle is really a circle. Kant regarded such diagrams as of only secondary psychological help to us. But he thought that our power of imagination—what he called the *Anschauung*, the intuition— was flawless. If we saw a geometrical truth clearly in our mind, not just with our eyes, then we saw it with complete certainty.

How would we approach, as a Kantian, the statement that two lines cannot have more than one point in common? We picture the situation in our mind. Here are two lines that cross at one point. How could they possibly cross somewhere else? Obviously, they cannot, because the lines get farther and farther apart as we move away from the crossing. It seems, therefore, quite clear that two lines either have all their points in common (in which case they coincide to become a single line), or they have, at most, one point or, possibly, no point in common. These simple truths of geometry, Kant said, we *see* immediately. We grasp their truth intuitively. The fact that we do not have to rely on diagrams led Kant to suppose that we can have complete confidence in truths perceived in this intuitive way. Later we shall come back to this view. We mention it here only to help the reader understand the way in which scientists at the beginning of the nineteenth century thought about geometry. Even if they had never read Kant, they had the same view. Whether their view derived from Kant or was just a part of the cultural atmosphere that Kant made explicit does not matter. Everyone assumed there were clear, simple, basic truths of geometry that were beyond doubt. From these simple truths, the axioms of geometry, one could pass, step by step, to certain derived truths, the theorems.

As we have said, some mathematicians believed that they could derive the parallel axiom from the other axioms of Euclid. Why were the flaws in their proofs so difficult to detect? The answer lies in the fact that, at the time, there did not exist a logic sufficiently powerful to provide strictly logical rules for geometrical proofs. At some place in the derivation an appeal to the imagination crept in, sometimes quite explicitly, sometimes in a hidden way. A method for distinguishing between a purely logical derivation and a derivation that brings in nonlogical components based on intuition became available only after the development of a systematized logic in the second half of the last century. The fact that this new logic was formulated in symbols increased its efficiency, but it was not absolutely essential. What was essential was, first, that the rules could be stated with complete exactness, and, second, that throughout the entire derivation no statement was made that could not be obtained from the premises or from previously obtained results by an application of the logic's rules of inference.

Before the development of modern logic, no system of logic existed with a set of rules adequate to cope with geometry. Traditional logic dealt only with one-place predicates, but in geometry we deal with relationships among many elements. A point lying on a line or a line lying on a plane are examples of two-place relations; a point lying between two other points is a three-place relation. We might think of congruence between two line segments as a two-place relation, but, since it is not customary to take line segments as primitive entities, a segment is best represented as a pair of points. In this case, congruence between two line segments is a relation between one point-pair and another point-pair; in other words, it is a four-place relation between points. As you see, geometry needs a logic of relations. This logic was not in existence at the time we are considering. When it became available, the logical flaws in various supposed proofs of the parallel axiom were revealed. At some point in each argument, an appeal was made to a premiss that rested on intuition and could not be derived logically from Euclid's other axioms. This might have been interesting, except for the fact that the hidden, intuitive premiss turned out, in every case, to be the parallel axiom itself in disguised form.

An example of such a disguised axiom, equivalent to the parallel axiom, is the following: If, in a plane, there is a straight line *L* and a curve *M*, if all points of *M* are the same distance from *L*, then *M* is also a straight line. This is shown in Figure 13–1, where *a* is the con-

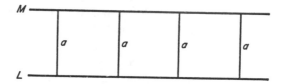

Figure 13–1.

stant distance from *L* of all points on *M*. This axiom, which intuitively seems true, was sometimes taken as a tacit assumption in attempts to prove the parallel axiom. When this is assumed, the parallel axiom can indeed be proved. Unfortunately, the assumption itself cannot be demonstrated unless we assume the truth of the parallel axiom or of some other axiom equivalent to it.

Another axiom equivalent to the parallel axiom, though perhaps not so intuitively obvious as the one just cited, is the assumption that geometrical figures of different sizes may be similar. Two triangles, for example, are said to be similar if they have equal angles and sides in the same proportion. In Figure 13–2, the ratio $a : b$ equals the ratio $a' : b'$, and the ratio $b : c$ equals the ratio $b' : c'$. Suppose I draw first only the smaller triangle with sides a, b, c. Is there a larger triangle with these same angles and with sides a', b', c' that are in the same

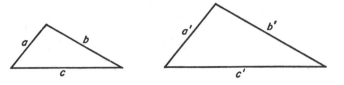

Figure 13–2.

proportion as *a, b, c* ? It seems obvious that the answer is yes. Suppose we wish to construct the larger triangle so that its sides are exactly twice as long as the sides of the smaller one. We can do it easily, as shown in Figure 13–3. We simply prolong side *a* by another segment

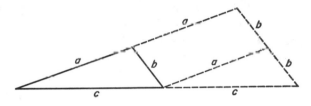

Figure 13–3.

of the same length, do the same to side *c*, then connect the end points. After giving it some thought, it seems quite clear that the third side must have a length of 2*b* and that the large triangle will be similar to the small one. If we assume this axiom about similar triangles, we can then prove the parallel axiom; but, once again, we are assuming the parallel axiom in disguised form. The truth is that we cannot prove the similarity of the two triangles without employing the parallel axiom or another one equivalent to it. To use the axiom about the triangles, therefore, is equivalent to using the parallel axiom, the very axiom we are trying to establish.

Not until the nineteenth century was it actually shown, by rigorous logic, that the parallel axiom is independent of the other axioms of Euclid. It cannot be derived from them. Negative statements, such as this, asserting the impossibility of doing something, are usually much harder to prove than positive statements. A positive statement that this or that *can* be derived from certain premises is demonstrated simply by showing the logical steps of the derivation. But how is it possible to prove that something is *not* derivable? If you fail in hundreds of attempts to derive it, you might give up, but that is not a proof of impossibility. It may be that someone else, perhaps in some unsuspected, roundabout way, will find a derivation. Nevertheless, difficult though it was, a formal proof of the independence of the parallel axiom was finally obtained.

Exploring the consequences of this discovery proved to be one of the most exciting developments in nineteenth-century mathematics. If the parallel axiom is independent of the other axioms of Euclid, then a statement incompatible with the parallel axiom can be substituted for it without logically contradicting the other axioms. By trying different alternatives, new axiom systems, called non-Euclidean geometries, were created. What was one to think of these strange new systems, with theorems so contrary to intuition? Should they be regarded as nothing more than a harmless logical game, a playing around with statements to see how they can be combined without logical inconsistency? Or should they be regarded as possibly "true" in the sense that they might apply to the structure of space itself?

This last case seemed so absurd, at the time, that no one had dreamed of even raising the question. In fact, when a few daring mathematicians began to study non-Euclidean systems, they hesitated to publish their investigations. One may laugh about it now and ask why feelings should be aroused by the publication of any system of mathematics. Today, we often take a purely formalistic approach to an axiom system. We do not ask what interpretations or applications it may have, but only whether the system of axioms is logically consistent and whether a certain statement is derivable from it. But this was not the attitude of most mathematicians in the nineteenth century. For them, a "point" in a geometrical system meant a position in the space of nature; a "straight line" in the system meant a straight line in the ordinary sense. Geometry was not viewed as an exercise in logic; it was an investigation of the space we find around us, not space in the abstract sense that mathematicians mean today when they speak about a topological space, a metric space, a five-dimensional space, and so on.

Carl Friedrich Gauss, one of the greatest mathematicians, perhaps *the* greatest mathematician, of the nineteenth century was the first, as far as anyone knows, to discover a consistent system of geometry in which the parallel axiom was replaced by an axiom inconsistent with it. We know this not from any publication of his, but only from a letter he wrote to a friend. In this letter, he speaks of studying such a system and deriving some interesting theorems from it. He adds that he did not care to publish those results because he was afraid of "the outcry of the Boeotians". The reader may know that, in ancient Greece, the Boeotians, inhabitants of the province of Boeotia, were not highly regarded. We can translate his statement into modern idiom by saying, "these hillbillies will laugh and say that I am crazy". By "hillbillies", however, Gauss did not mean unlearned people; he meant certain professors of mathematics and philosophy. He knew they would think him out of his mind to be taking a non-Euclidean geometry seriously.

If we give up the parallel axiom, what can be put in its place? The answer to this question, one of the most important questions in the history of modern physics, will be considered in detail in chapters 14 through 17.

CHAPTER 14

Non-Euclidean Geometries

IN SEARCHING FOR an axiom to put in place of Euclid's parallel axiom, there are two opposite directions in which we can move:

(1) We can say that there is *more than one* parallel. (It turns out that, if there is more than one, there will be an infinite number.)

(2) We can say that on a plane, through a point outside a line, there is *no* parallel. (Euclid had said there is exactly one.)

The first of these deviations from Euclid was explored by the Russian mathematician Nikolai Lobachevski, the second by the German mathematician Georg Friedrich Riemann. In the chart in Figure 14–1, I have placed the two non-Euclidean geometries on opposite sides of the Euclidean to emphasize how they deviate from the Euclidean structure in opposite directions.

Lobachevski's geometry was discovered independently and almost simultaneously by Lobachevski, who published his work in 1835, and by the Hungarian mathematician Johann Bolyai, who published his results three years earlier. Riemann's geometry was not discovered until about twenty years later. If you would like to look further into the

type of geometry	number of parallels	sum of angles in triangle	ratio of circumference to diameter of circle	measure of curvature
Lobachevski	∞	<180°	>π	<0
Euclid	1	180°	π	0
Riemann	0	>180°	<π	>0

Figure 14–1.

subject of non-Euclidean geometries, there are several good books available in English. One is *Non-Euclidean Geometry* by the Italian mathematician Roberto Bonola. It contains the two articles by Bolyai and Lobachevski, and it is interesting to read them in their original form. I think the best book that discusses non-Euclidean geometry from the point of view adopted here, namely, its relevance to the philosophy of geometry and space, is Hans Reichenbach's *Philosophie der Raum-Zeit-Lehre,* first published in 1928 but now available in English translation as *The Philosophy of Space and Time.* If you are interested in the historical point of view, there is Max Jammer's book, *Concepts of Space: The History of Theories of Space in Physics.* Sometimes Jammer's discussions are a bit metaphysical. I am not sure whether this is due to his own views or to those of the men he is discussing; in any case, it is one of the few books that takes up in detail the historical development of the philosophy of space.

Let us look more closely at the two non-Euclidean geometries. In the Lobachevski geometry, technically called hyperbolic geometry, there are an infinite number of parallels. In the Riemann geometry, known as elliptic geometry, there are no parallels. How is a geometry that does not contain parallel lines possible? We can understand this by turning to a model that is not exactly the model of an elliptic geometry, but one closely related to it—a model of spherical geometry. The model is simply the surface of a sphere. We view this surface as analogous to a plane. Straight lines on a plane are here represented by the great circles of the sphere. In more general terms, we say that in any non-Euclidean geometry the lines that correspond to straight lines in Euclidean geometry are "geodesic lines". They share with straight lines the property of being the shortest distance between two given points. On our model, the surface of the sphere, the shortest distance between two points, the geodesic, is a portion of a great circle. Great circles are the curves obtained by cutting the sphere with a plane through the sphere's center. The equator and the meridians of the earth are familiar examples.

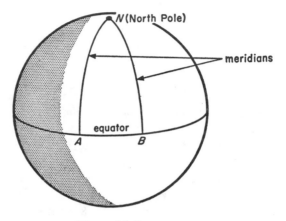

Figure 14–2.

In Figure 14–2 two meridians have been drawn perpendicular to the equator. In Euclidean geometry, we expect two lines perpendicular to a given line to be parallel, but on the sphere these lines meet at the North Pole and also at the South Pole. On the sphere there are no two straight lines, or, rather, quasistraight lines, *i.e.,* great circles, that do not meet. We have here, then, an easily imaginable model of a geometry in which there are no parallel lines.

The two non-Euclidean geometries can also be distinguished by the sum of the angles of a triangle. This distinction is important from the standpoint of empirical investigations of the structure of space. Gauss was the first to see clearly that only an empirical investigation of space can disclose the nature of the geometry that best describes it. Once we realize that non-Euclidean geometries can be logically consistent, we can no longer say, without making empirical tests, which geometry holds in nature. In spite of the Kantian prejudice prevailing in his time, Gauss may actually have undertaken an experiment of this sort.

It is easy to see that testing triangles is much easier than testing parallel lines. Lines thought to be parallel might not meet until they had been prolonged for many billions of miles, but measuring the angles of a triangle can be undertaken in a small region of space. In Euclidean geometry the sum of the angles of any triangle is equal to two right angles, or 180 degrees. In Lobachevski's hyperbolic geometry, the sum of the angles of any triangle is less than 180 degrees. In the Riemannian elliptic geometry the sum is greater than 180 degrees.

The deviation from 180 degrees, in elliptic geometry, is easily understood with the aid of our model, the surface of a sphere. Consider the triangle NAB in Figure 14–2; it is formed by segments of two meridians and the equator. The two angles at the equator are 90 degrees, so we already have a total of 180 degrees. Adding the angle at the North Pole will bring the sum to more than 180. If we move the meridians until they cross each other at right angles, each angle of the triangle will be a right angle, and the sum of all three will be 270 degrees.

We know that Gauss thought of making a test of the sum of the angles of an enormous stellar triangle, and there are reports that he actually carried out a similar test, on a terrestrial scale, by triangulating three mountain tops in Germany. He was a professor at Göttingen, so it is said that he chose a hill near the city and two mountain tops that could be seen from the top of this hill. He had already done important work in applying the theory of probability to errors of measurement, and this would have provided an opportunity to make use of such procedures. The first step would have been to measure the angles optically from each summit, repeating the measurement many times. By taking the mean of these observational results, under certain constraints, he could determine the most probable size of each angle and, therefore, the most probable value for their sum. From the dispersion of the results, he could then calculate the probable error; that is, a certain interval around the mean, such that the probability of the true value lying within the interval was equal to the probability of it lying outside the interval. It is said that Gauss did this and that he found the sum of the three angles to be not exactly 180 degrees, but deviating by such a small amount that it was within the interval of probable error. Such a result would indicate either that space is Euclidean or, if non-Euclidean, that its deviation is extremely small—less than the probable error of the measurements.

Even if Gauss did not actually make such a test, as recent scholarship has indicated, the legend itself is an important milestone in the history of scientific methodology. Gauss was certainly the first to ask the revolutionary question, what shall we find if we make an empirical investigation of the geometrical structure of space? No one else had thought of making such an investigation. Indeed, it was considered preposterous, like trying to find by empirical means the product of seven and eight. Imagine that we have here seven baskets, each containing eight balls. We count all the balls many times. Most of the time we get 56, but occasionally we get 57 or 55. We take the mean of these results to discover the true value of seven times eight. The French mathematician P. E. B. Jourdain once jokingly suggested that the best way to do this would be not to do the counting yourself, because you are not an expert in counting. The experts are the headwaiters, who are constantly adding and multiplying numbers. The most experienced headwaiters should be brought together and asked how much seven times eight is. One would not expect much deviation in their answers, but if you use larger numbers, say, 23 times 27, there would be some dispersion. We take the mean of all their answers, weighted according to the number of waiters who gave each answer, and, on this basis, we obtain a scientific estimate of the product of 23 and 27.

Any attempt to investigate empirically a geometrical theorem seemed just as preposterous as this to Gauss's contemporaries. They viewed geometry in the same way they viewed arithmetic. They believed, with Kant, that our intuition does not make geometrical mistakes. When we "see" something in our imagination, it cannot be otherwise. That someone should measure the angles of a triangle—not just for fun or to test the quality of optical instruments, but to find the true value of their sum—seemed entirely absurd. Everyone could see, after a little training in Euclidean geometry, that the sum *must* be 180 degrees. For this reason, it is said, Gauss did not publish the fact that he made such an experiment, nor even that he regarded such an experiment as worth doing. Nevertheless, as a result of continued speculation about non-Euclidean geometries, many mathematicians began to realize that these strange new geometries posed a genuine empirical problem. Gauss himself did not find a conclusive answer; but he provided a strong stimulation for thinking in a non-Kantian way about the whole problem of the structure of space in nature.

To see more clearly how the various non-Euclidean geometries differ from one another, let us again consider the surface of a sphere. As we have seen, this is a convenient model that helps us understand intuitively the geometrical structure of a plane in Riemannian space. (Riemannian space here means what is called elliptical space. The term

"Riemannian space" also has a more general meaning that will be clarified later.)

We must be careful not to overextend the analogy between the Riemannian plane and the sphere's surface, because any two straight lines on a plane in Riemannian space have only one point in common, whereas the lines on a sphere that correspond to straight lines—the great circles—always meet at *two* points. Consider, for example, two meridians. They meet at both the North Pole and the South Pole. Strictly speaking, our model corresponds to the Riemannian plane only if we restrict ourselves to a portion of the sphere's surface that does not contain opposite points, like the North and South poles. If the entire sphere is our model, we must assume that each point on the Riemannian plane is represented on the surface of the sphere by a pair of opposite points. Starting from the North Pole and traveling to the South Pole on the earth would correspond to starting from one point on the Riemannian plane, traveling in a straight line on the plane, and returning to that same point. All geodesic lines in Riemannian space have the same finite length and are closed, like the circumference of a circle. The extreme deviation of this fact from our intuition is probably the reason this kind of geometry was discovered later than Lobachevski's geometry.

With the aid of our spherical model, we easily see that, in Riemannian space, the ratio of a circle's circumference to its diameter is always less than pi. Figure 14–3 shows a circle on the earth that has the

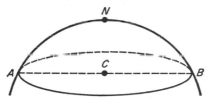

Figure 14–3.

North Pole for its center. This corresponds to a circle in the Riemannian plane. Its radius is not the line *CB*, because that does not lie on the sphere's surface, which is our model. The radius is the arc *NB*, and the diameter is the arc *ANB*. We know that the circumference of this circle has the ratio of pi to the line segment *ACB*. Since the arc *ANB* is longer than the segment *ACB*, it is clear that the ratio of the circle's perimeter to *ANB* (the circle's diameter in the Riemannian plane) must be less than pi.

It is not so easy to see that in the Lobachevski space it is just the other way: the ratio of a circle's circumference to its diameter must be greater than pi. Perhaps we can visualize it with the aid of another model. This model (shown in Figure 14–4) cannot be used for the

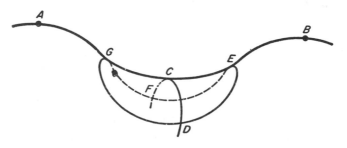

Figure 14–4.

entire Lobachevski plane—certainly not for three-dimensional Lobachevski space—but it can be used for a limited portion of the Lobachevski plane. The model is a saddle-shaped surface resembling a pass between two mountains. *A* is one mountain top, *C* is the pass, *B* is the other mountain top. Try to visualize this surface. There is a curve, perhaps a path, passing through point *F* on the far side of the pass, rising over the pass through point *C,* then going down on the near side of the pass through point *D.* The saddle-shaped portion of this surface, including points *C, D, E, F, G,* can be regarded as a model of the structure in a Lobachevski plane.

What form does a circle have on this model? Assume that the center of a circle is at *C.* The curved line *DEFGD* represents the circumference of a circle that is at all points the same distance from the center *C.* If you stand at point *D,* you find yourself lower than the circle's center; if you walk along the circle to *E,* you find yourself higher than the center. It is not hard to see that this wavy line, which corresponds to a circle in the Lobachevski plane, must be longer than an ordinary circle on a Euclidean plane that has *CD* for its radius. Because it is longer, the ratio of the circumference of this circle to its diameter (arc *FCD* or arc *GCE*) must be greater than pi.

A more exact model, corresponding accurately in all measurements. to a part of a Lobachevski plane, can be constructed by taking a cer-

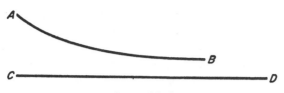

Figure 14-5.

tain curve, called a tractrix (arc *AB* in Figure 14–5), and rotating it around the axis *CD*. The surface generated by this rotation is called a pseudosphere. Perhaps you have seen a plaster of Paris model of this surface. If you study such a model, you can see that triangles on its surface have three angles totaling less than 180 degrees and that circles have a ratio of circumference to diameter that exceeds pi. The larger the circle on such a surface, the greater will be the ratio's deviation from pi. We must not think of this as meaning that pi is not a constant. Pi is the ratio of the circumference of a circle in a Euclidean plane to its diameter. This fact is not altered by the existence of non-Euclidean geometries in which the ratio of a circle's circumference to its diameter is a variable that may be greater or less than pi.

All surfaces, both Euclidean and non-Euclidean, have at any of their points a measure called the "measure of curvature" of that surface at that point. The Lobachevski geometry is characterized by the fact that, in any plane, at any point, the plane's measure of curvature is negative and constant. There is an infinite number of different Lobachevski geometries. Each is characterized by a certain fixed parameter—a negative number—that is the measure of curvature of a plane in that geometry.

You might object that, if it is a plane, then it cannot have a curvature. But "curvature" is a technical term and is not to be understood here in the ordinary sense. In Euclidean geometry we measure the curvature of a line at any point by taking the reciprocal of its "radius of curvature". "Radius of curvature" means the radius of a certain circle that coincides, so to speak, with an infinitesimal part of the line at the point in question. If a curved line is almost straight, the radius of curvature is long. If the line is strongly curved, the radius is short.

How do we measure the curvature of a surface at a given point? We first measure the curvature of two geodesics that intersect at that point and extend in two directions, called the "principal directions" of the surface at that point. One direction gives the maximum curvature of a geodesic at that point, and the other gives the minimum curvature. We then define the curvature of the surface at that point as the product of the two reciprocals of the two radii of curvature of the two geodesics. For example, consider the mountain pass shown in Figure 14–4. How do we measure the curvature of this surface at point C? We see that one geodesic, the arc GCE, curves in a concave manner (looking down on the surface), whereas the geodesic at right angles to it, arc FCD, curves in a convex manner. These two geodesics give the maximum and minimum curvatures of the surface at point C. Of course, if we look *up* at this surface from the underside, arc GCE appears convex, and arc FCD appears concave. It does not matter at all from which side we view the surface, which curve we wish to consider convex and which concave. By convention, we call one side positive and the other negative. The product of the reciprocals of these two radii, $\dfrac{1}{R_1 R_2}$, gives us the measure of curvature of the saddle surface at point C. At any point on the saddle surface, one radius of curvature will be positive, the other negative. The product of the two reciprocals of those radii and, consequently, the measure of curvature of the surface, must therefore always be negative.

This is not the case with respect to a surface that is completely convex, such as that of a sphere or an egg. On such a surface, the two geodesics, in the two principal directions, both curve the same way. One geodesic may curve more strongly than the other, but both curve in the same manner. Again, it does not matter whether we view such a surface from one side and call the two radii of curvature positive or from the other and call them negative. The product of their reciprocals will always be positive. Therefore, on any convex surface such as that of a sphere, the measure of curvature at any point will be positive.

The Lobachevski geometry, represented by the saddle-surface model, can be characterized in this way: for any Lobachevski space, there is a certain negative value that is the measure of curvature for any point in any plane in that space. The Riemannian geometry, represented by the spherical surface, can be characterized in a similar way: for any Riemannian space, there is a certain positive value that is the measure of curvature for any point on any plane in that space. Both are spaces of constant curvature. This means that, for any one such space, the measure of curvature at any point, in any plane, is the same.

Let k be the measure of curvature. In Euclidean space, which also has a constant curvature, $k = 0$. In Lobachevski space, $k < 0$. In

Riemannian space, $k > 0$. These numerical values are not determined by the axioms of the geometry. Different Riemannian spaces are obtained by choosing different positive values for k and different Lobachevski spaces are obtained by choosing different negative values for k. Aside from the value of the parameter k, all the theorems are entirely alike in all Lobachevski spaces and are entirely alike in all Riemannian spaces. Of course, the theorems of each geometry are quite different from those of the other.

It is important to realize that "curvature", in its original and literal sense, applies only to surfaces of a *Euclidean model* of a non-Euclidean plane. The sphere and the pseudosphere are curved surfaces in this sense. But the term "measure of curvature", as applied to non-Euclidean planes, does not mean that these planes "curve" in the ordinary sense. Generalizing the term "curvature", so that it applies to non-Euclidean planes, is justified, because the internal geometrical structure of a Riemannian plane is the same as the structure of the surface of a Euclidean sphere; the same is true of the structure of the plane in Lobachevski space and the surface of a Euclidean pseudosphere. Scientists often take an old term and give it a more general meaning. This caused no difficulty at all during the nineteenth century, because non-Euclidean geometries were studied only by mathematicians. The trouble began when Einstein made use of non-Euclidean geometry in his general theory of relativity. This took the subject out of the field of pure mathematics and into the field of physics, where it became a description of the actual world. People wanted to understand what Einstein was doing, so books were written explaining these things to the layman. In those books, the authors sometimes discussed "curved planes" and "curved space". That was an extremely unfortunate, misleading way of speaking. They should have said: "There is a certain measure k—mathematicians call it 'measure of curvature', but don't pay any attention to that phrase—and this k is positive inside the sun but negative in the sun's gravitational field. As we go farther away from the sun, the negative value of k approaches zero."

Instead of putting it this way, popular writers said that Einstein had discovered that the planes in our space are curved. That could only confuse the layman. Readers asked what it means to say that planes are curved. If they are curved, they thought, they should not be called planes! Such talk of curved space led people to believe that everything in space is distorted, or bent. Sometimes the writers of books on relativity even talked about how the force of gravitation bends the planes. They described it with real feeling, as if it were analogous to someone bending a metal sheet. This type of thinking led to strange consequences, and some writers objected to Einstein's theory on those grounds. All this could have been avoided if the term "curvature" had been avoided.

On the other hand, to introduce a term entirely different from one already in customary use in mathematics is not easy to do. The best procedure, therefore, is to accept the term "curvature" as a technical term but clearly understand that this term should not be connected with the old associations. Do not think of a non-Euclidean plane as being "bent" into a shape that is no longer a plane. It does not have the internal structure of a Euclidean plane, but it is a plane in the sense that the structure on one side of it is exactly like the structure on the other side. Here we see the danger in saying that the Euclidean sphere is a model of the Riemannian plane, because, if you think of a sphere, you think of the inside as quite different from the outside. From the inside, the surface looks concave; from the outside, it is convex. This is not true of the plane in either the Lobachevski or Riemannian space. In both spaces the two sides of the plane are identical. If we leave the plane on one side, we observe nothing different from what we observe if we leave the plane on the other side. But the inner structure of the plane is such that we can, with the help of the parameter k, measure its degree of "curvature". We must remember that this is curvature in a technical sense, and is not quite the same as our intuitive understanding of curvature in Euclidean space.

Another terminological confusion, easily cleared up, concerns the two meanings (we alluded to them earlier in this chapter) of "Riemannian geometry". When Riemann first devised his geometry of constant positive curvature, it was called Riemannian to distinguish it from the earlier space of Lobachevski, in which the constant curvature is negative. Later, Riemann developed a generalized theory of spaces with variable curvature, spaces that have not been dealt with axiomatically. (The axiomatic forms of non-Euclidean geometry, in which all of Euclid's axioms are retained except that the parallel axiom has been replaced by a new axiom, are confined to spaces of constant curvature.) In Riemann's general theory, any number of dimensions can be considered, and, in all cases, the curvature may vary continuously from point to point.

When physicists speak of "Riemannian geometry", they mean the generalized geometry in which the old Riemannian and Lobachevski geometries (today called elliptic and hyperbolic geometries), together with Euclidean geometry, are the simplest special cases. In addition to those special cases, generalized Riemannian geometry contains a great variety of spaces of varying curvature. Among these spaces is the space Einstein adopted for his general theory of relativity.

CHAPTER 15

Poincaré versus
Einstein

HENRI POINCARÉ, a famous French mathematician and physicist and the author of many books on the philosophy of science, most of them before the time of Einstein, devoted much attention to the problem of the geometrical structure of space. One of his important insights is so essential for an understanding of modern physics that it will be worthwhile to discuss it in some detail.[1]

Suppose, Poincaré wrote, that physicists should discover that the structure of actual space deviated from Euclidean geometry. Physicists would then have to choose between two alternatives. They could either accept non-Euclidean geometry as a description of physical space, or they could preserve Euclidean geometry by adopting new laws stating that all solid bodies undergo certain contractions and expansions. As we have seen in earlier chapters, in order to measure accurately with a steel rod, we must make corrections that account for the thermal expansions or contractions of the rod. In a similar way, said Poincaré, if observations suggested that space was non-Euclidean, physicists could retain

[1] Poincaré's view on this matter is brought out most explicitly in his *Science and Hypothesis* (London: 1905); (New York: Dover, 1952).

144

Euclidean space by introducing into their theories new forces—forces that would, under specified conditions, expand or contract solid bodies.

New laws would also have to be introduced in the field of optics, because we can also study physical geometry by means of light rays. Such rays are assumed to be straight lines. The reader will recall that the three sides of Gauss's triangle, which had mountains for vertices, did not consist of solid rods—the distances were much too great—but of light rays. Suppose, Poincaré said, that the sum of the angles of a large triangle of this sort were found to deviate from 180 degrees. Instead of abandoning Euclidean geometry, we could say that the deviation is due to a bending of light rays. If we introduce new laws for the deflection of light rays, we can always do it in such a way that we keep Euclidean geometry.

This was an extremely important insight. Later, I shall try to explain just how Poincaré meant it and how it can be justified. In addition to this far-reaching insight, Poincaré predicted that physicists would always choose the second way. They will prefer, he said, to keep Euclidean geometry, because it is much simpler than non-Euclidean. He did not know, of course, of the complex non-Euclidean space that Einstein would soon propose. He probably thought only of the simpler non-Euclidean spaces of constant curvature; otherwise, he would no doubt have thought it even *less* likely that physicists would abandon Euclid. To make a few alterations in the laws that concern solid bodies and light rays seemed, to Poincaré, justified on the ground that it would retain the simpler system of Euclid. Ironically, it was just a few years later, in 1915, that Einstein developed his general theory of relativity, in which non-Euclidean geometry was adopted.

It is important to understand Poincaré's point of view; it helps us to understand Einstein's reasons for abandoning it. We will try to make it clear in an intuitive way, rather than by calculations and formulas, so that we can visualize it. To do this, we will use a device employed by Hermann von Helmholtz, the great German physicist, many decades before Poincaré wrote on the topic. Helmholtz wanted to show that Gauss had been right in regarding the geometrical structure of space as an empirical problem. Let us imagine, he said, a two-dimensional world in which two-dimensional beings walk about and push around objects. These beings and all the objects in their world are completely flat, like the two-dimensional creatures in Edwin A. Abbott's amusing fantasy, *Flatland*. They live, not on a plane, but on the surface of a sphere. The sphere is gigantic in relation to their own size; they are the size of ants, and the sphere is as large as the earth. It is so large that they never travel all the way around it. In other words, their movements are confined to a limited domain on the surface of the sphere. The question is, can these creatures, by making internal measurements on their two-dimensional surface, ever discover whether they are on a plane or a sphere or some other kind of surface?

Helmholtz answered that they can. They could make a very large triangle and measure the angles. If the sum of the angles were greater than 180 degrees, they would know they were on a surface with positive curvature; if they found the same positive curvature at every point on their continent, they would know they were on the surface of a sphere or of part of a sphere. (Whether the sphere is complete or not is another question.) The hypothesis that their whole universe was a spherical surface would be reasonable. We, of course, can see at a glance that it is such a surface because we are three-dimensional creatures who stand outside it. But Helmholtz made it clear that the two-dimensional creatures themselves, by measuring the angles of a triangle or the ratio of the circle to its diameter (or various other quantities), could calculate the measure of curvature at each spot on their surface. Gauss was right, therefore, to think he could determine whether our three-dimensional space has a positive or negative curvature by making measurements. If we imagine our space imbedded in a higher-dimensional universe, we can speak of a real bend or curvature of our space, for it would appear curved to four-dimensional creatures.

We must examine this a little more closely. Suppose that the two-dimensional creatures discover that, when they measure triangles with their measuring rods, at every point on their continent there is the same positive curvature for triangles of the same size. Among these creatures are two physicists, P_1 and P_2. Physicist P_1 maintains theory T_1, which says that the region on which he and his fellow-creatures live is part of a spherical surface S_1. His colleague, physicist P_2, maintains theory T_2, which says that the region is a flat surface S_2. In Figure 15–1 these two surfaces are drawn in profile. Let us assume that in S_1 there are rigid two-dimensional bodies, such as creatures and measuring rods, that move about without change of size or shape. For every body in S_1 there is a corresponding flat body in S_2, which is its projection, a projection made by, say, parallel lines perpendicular to the plane S_2 (in the illustration these parallel lines are shown as broken lines). If a body in

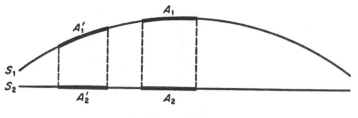

Figure 15–1.

S_1 moves from position A_1 to A_1', its shadow body in S_2 moves from A_2 to A_2'. We assume that bodies in S_1 are rigid; therefore, the length A_1 is equal to that of A_1'. But this means that A_2' must be shorter than A_2.

Helmholtz pointed out that, when we measure something with a measuring rod, what we actually observe is nothing more than a series of point coincidences. This can easily be seen from our earlier description of the measurement of the edge of a fence, at the beginning of Chapter 9.

Look once more at Figure 15–1. The projection from S_1 to S_2 is called a one-to-one mapping. (This could not be done if S_1 were an entire sphere, but we have assumed that S_1 is only a limited region on a sphere.) For every point on S_1, there is exactly one corresponding point on S_2. Therefore, as beings move about on S_1, observing point coincidences between their measuring rods and what they are measuring, their shadow beings on S_2 make exactly the same observations on the corresponding shadow bodies. Since the bodies in S_1 are assumed to be rigid, the corresponding bodies in S_2 cannot be rigid. They must suffer certain contractions and expansions such as we have indicated in the illustration.

Let us return to the two physicists, P_1 and P_2, who hold different theories about the nature of their flat world. P_1 says that this world must be part of a sphere. P_2 insists that it is a plane but that bodies expand and contract in certain predictable ways as they move around. For example, they get longer as they move toward the central part of S_2, shorter as they move away from the center. P_1 maintains that light rays are geodesics on the curved surface S_1; that is, they follow the arcs of great circles. These arcs will project to S_2 as the arcs of ellipses. P_2, in order to defend his theory that the world is a plane, must, therefore, devise optical theories in which light rays move in elliptical paths.

How can the two physicists decide which of them is right? The answer is that there is no way of deciding. Physicist P_1 contends that their world is part of the surface of a sphere and that bodies do not suffer contractions and expansions except, of course, for such familiar phenomena (or, rather, the two-dimensional analogs of such phenomena) as thermal expansion, elastic expansion, and so on. Physicist P_2 describes the same world in a different way. He thinks it is a plane but that bodies expand and contract in certain ways as they move over the surface. We, who are in a three-dimensional space, can observe this two-dimensional world and see whether it is a sphere or plane, but the two physicists are restricted to their world. They cannot in principle decide which theory is correct. For this reason, Poincaré said, we should not even raise the question of who is right. The two theories are no more than two different methods of describing the same world.

There is an infinity of different ways that physicists on the sphere could describe their world, and, according to Poincaré, it is entirely a matter of convention which way they choose. A third physicist might hold the fantastic theory that the world had this shape:

He could defend such a theory by introducing still more complicated laws of mechanics and optics, laws that would make all observations compatible with the theory. For practical reasons, no physicist on the sphere would wish to propose such a theory. But, Poincaré insisted, there is no logical reason why he could not do so.

We can imagine a two-dimensional analog of Poincaré saying to the rival physicists: "There is no need to quarrel. You are simply giving different descriptions of the same totality of facts." Leibniz, the reader may recall, had earlier defended a similar point of view. If there is in principle no way of deciding between two statements, Leibniz declared, we should not say they have different meanings. If all bodies in the universe doubled in size overnight, would the world seem strange to us next morning? Leibniz said it would not. The size of our own bodies would double, so there would be no means by which we could detect a change. Similarly, if the entire universe moved to one side by a distance of ten miles, we could not detect it. To assert that such a change had occurred would, therefore, be meaningless. Poincaré adopted this view of Leibniz's and applied it to the geometrical structure of space. We may find experimental evidence suggesting that physical space is non-Euclidean, but we can always keep the simpler Euclidean space if we are willing to pay a price for it. As we have seen, Poincaré did not think that this price would ever be too high.

There are two basic points that our consideration of the flat world was intended to make clear and that we shall apply to our actual world. First, by making use of ordinary measuring procedures to which we are accustomed, we might arrive at the result that space has a non-Euclidean structure. Some recent philosophers (Hugo Dingler, for example) have not been able to see this. They hold that our measuring procedures employ instruments that have been manufactured under the assumption that geometry is Euclidean; therefore, these instruments could not possibly give us anything but Euclidean results. This contention is certainly wrong. Our instruments occupy such tiny parts of space that the question of how our space deviates from Euclidean geometry does not enter into their construction. Consider, for example, a surveyor's instrument for measuring angles. It contains a circle divided into 360 equal parts, but it is such a small circle that, even if space deviated from the Euclidean to a degree that Gauss hoped he could measure (a much greater degree than the deviation in relativity theory), it would still have no effect on the construction of this circle. In small regions of space, Euclidean geometry would still hold with very high approximation. This is sometimes expressed by saying that non-Euclidean space has a Euclidean structure in small environments. From a strict mathematical standpoint, it is a matter of a limit. The smaller the region of space, the closer its structure gets to the Euclidean. But our laboratory instruments occupy such minute portions of space that we can completely disregard any influence non-Euclidean space might have on their construction.

Even if the deviation from Euclidean geometry were so strong that the sum of the angles in a small triangle (say, one drawn on a designer's board) would differ considerably from 180 degrees, that fact could certainly be determined with the help of instruments made in the customary way. Suppose that the beings on the spherical surface S_1 (see Figure 15–1) construct a protractor by cutting a circular disk and dividing its circumference into 360 equal parts. If this protractor were used for measuring the angles of a triangle formed (as in an earlier example) by two half meridians and a quarter of the equator, it would show each angle to be 90 degrees and, therefore, the sum of the three angles to be 270 degrees.

The second basic point brought out by our consideration of the two-dimensional world is that, if we find empirical evidence of a non-Euclidean space, we can preserve Euclidean geometry provided we are willing to introduce complications into the laws that govern solid bodies and the laws of light rays. When we look at surfaces within our space, such as a surface on which we see an ant crawling, it is meaningful to ask whether the surface is a plane, or part of a sphere, or some other type of surface. On the other hand, if we are dealing with the space of our universe, a space we cannot observe as something imbedded in a universe of higher dimensions, then it is meaningless to ask whether space is non-Euclidean or whether our laws must be modified to preserve Euclidean geometry. The two theories are merely two descriptions of the same facts. We can call them equivalent descriptions because we make exactly the same predictions about observable events in both theories. Perhaps "observationally equivalent" would be a more appropriate phrase. The theories may differ considerably in their logical structure, but if their formulas and laws always lead to the same predictions about observable events, we can say that they are equivalent theories.

At this point, it is well to distinguish clearly between what we mean here by equivalent theories and what is sometimes meant by this phrase. Occasionally two physicists will propose two different theories to account for the same set of facts. Both theories may successfully explain this set of facts, but the theories may not be the same with respect to observations not yet made. That is, they may contain different predictions about what may be observed at some future time. Even though two such theories account completely for known observations, they should be regarded as essentially different physical theories.

Sometimes it is not easy to devise experiments that will distinguish between two rival theories that are not equivalent. A classic example is provided by Newton's theory of gravitation and Einstein's theory of gravitation. Differences in the predictions of these two theories are so small that clever experiments had to be devised and precise measurements made before it could be decided which theory made the best predictions. When Einstein later proposed his unified field theory, he said he was unable to think of any crucial experiment that could decide between this theory and other theories. He made it clear that his theory was not equivalent to any previous theory, but it was so abstractly stated that he was unable to deduce any consequences that could be observed under the present degree of precision of our best instruments. He believed that, if his unified field theory were investigated further or if our instruments improved sufficiently, it might be possible some day to make a decisive observation. It is very important to understand that "equivalent theories", as used here, means something much stronger than the fact that two theories account for all known observations. Equivalence here means that two theories lead in all cases to exactly the same predictions, like the theories of the two physicists in our flatland illustration.

In the next two chapters we will see in detail how Poincaré's insight into the observational equivalence of Euclidean and non-Euclidean theories of space leads to a deeper understanding of the structure of space in relativity theory.

CHAPTER 16

Space in
Relativity Theory

ACCORDING TO Einstein's theory of relativity, as discussed in previous chapters, space has a structure that deviates in gravitational fields from the structure of Euclidean geometry. Unless the gravitational field is extremely strong, the deviations are difficult to observe. The earth's gravitational field, for example, is so weak that it is not possible, even with the best instruments available, to detect any deviation from Euclidean structure in its vicinity. But, when much stronger gravitational fields, such as those surrounding the sun or stars with even larger masses than the sun, are considered, then certain deviations from Euclidean geometry are subject to observational testing.

The popular books that have been written about relativity theory as well as many other books in which the subject is discussed sometimes contain misleading statements. One page may state that Einstein's theory asserts that the structure of space in the gravitational field is non-Euclidean. On another page, or perhaps even on the same page, it is said that, according to relativity theory, rods contract in a gravitational field. (This is not the kind of contraction, sometimes called the Lorentz-contraction, that has to do with moving rods, but a contraction of rods at rest in a gravitational field.)

It must be made quite clear that these two statements do not fit together. It cannot be said that one is wrong. The author is right on one page. He is also right on the other page. But the two statements should not be on two pages in the same chapter. They belong to different languages, and the author should decide whether he wants to talk about relativity theory in one language or the other. If he wants to talk in Euclidean language, it is quite proper to speak of a rod contracting in a gravitational field. But he cannot also speak of a non-Euclidean structure of space. On the other hand, he may choose to adopt a non-Euclidean language; but then he cannot speak of contractions. Each language provides a legitimate way of talking about gravitational fields, but to mix the languages in the same chapter is very confusing to the reader.

It may be recalled that, in our previous discussion of the flat world, we imagined two physicists who held two different theories about the nature of their world. It became apparent that these two theories were really equivalent, differing only in that they were two different ways of describing the same totality of facts. The same situation holds with respect to relativity theory. One description, which we will call T_1, is non-Euclidean. The other, T_2, is Euclidean.

If the language of T_1, the non-Euclidean language, is chosen, the laws of mechanics and optics remain the same as in pre-Einsteinian physics. Solid bodies are rigid except for certain deformations, such as elastic expansions and contractions (when outside forces compress or stretch them), thermal expansions, changes produced by magnetization, and so forth. These deformations are a familiar part of classical physics and are taken care of by introducing various correcting factors into the definition of length. For example, it may be decided that a certain measuring rod will be the standard unit of length. Since it is known that the rod expands when it is heated, the rod represents this unit of length only when it has a certain "normal" temperature, T_0. Of course, the rod may at any given time have another temperature, T, that differs from T_0. Therefore, to define the length of the standard rod at temperature T, the normal length of the rod, l_0, must be multiplied by a correction factor, as explained in Chapter 9. In that chapter, this factor was expressed as $1 + \beta (T - T_0)$, where the value of β depends on the substance of the rod. Thus, the definition of length, l, is reached:

$$l = l_0 [1 + \beta(T - T_0)]$$

In similar fashion, other forces that may influence the rod's length must be taken into account, but gravity will not be among them. With respect to light, the language of T_1 asserts that light rays in a vacuum are always straight lines. They are not bent or deflected in any way by gravitational fields. The alternative description, T_2, preserves Euclidean geometry. Observations that suggest a non-Euclidean space are accounted for by modifications of the classical laws of optics and mechanics.

To see how these two descriptions can be applied to the structure of a plane in physical space, as conceived in Einstein's theory of relativity, consider a plane S passing through the center of the sun. According to relativity theory, observational tests (if feasible) would show that a triangle on this plane outside the sun would have angles totaling less than 180 degrees. Similarly, a circle on this plane, outside the sun, would have a ratio of circumference to diameter greater than pi. Measurements made *inside* the sun would show opposite deviations.

To make the structure of this plane intuitively clearer and to see how this structure can be described in the rival languages of T_1 and T_2, we make use of a model in Euclidean space that can be put into one-to-one correspondence with the structure of the non-Euclidean plane just described. This model is a certain curved surface, S', the construction of which is described here.[1]

In the coordinate system R-Z (see Figure 16–1), the curve DBC is an arc of a parabola that has Z for its directrix. (The curve is generated by a point that moves so that its perpendicular distance from the directrix is always the same as its distance from point F, the focus of the parabola.) V is the vertex of the parabola, and the distance a is proportional to the mass of the sun. The arc AB is the arc of a circle. Its center, E, is on the Z-axis, and it is placed so that the arc goes smoothly over into the parabola; this means that the tangent to the circle at B and the tangent to the parabola at B coincide. (B is called an inflection point of the curve ABC.) Suppose that this smooth curve ABC is rotated around the Z-axis to produce a surface similar to the surface of a hill. This is the surface S', which will serve as a Euclidean model of the non-Euclidean plane passing through the center of the sun.

The portion of the surface near the top of the hill, $B'AB$, is spherical and convex; it corresponds to the part of the plane inside the sun. Here the curvature is constant and positive. (This point is seldom made

---

[1] For this construction, see L. Flamm, *Physikalische Zeitschrift* (Leipzig), 17 (1916), 448–454, based on Karl Schwarzschild, *Sitzungsberichte der Preussischen Akademie der Wissenschaften* (Berlin: 1916), pp. 189–196, 424–434.

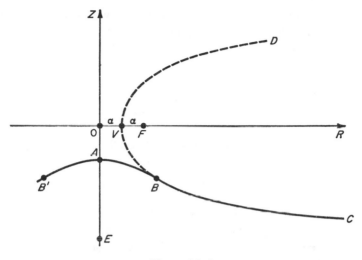

Figure 16–1.

in books on relativity theory, because few physicists are concerned with the geometrical structure of space inside a huge mass like the sun. But it is an important theoretical point and will be considered later, when a triangle of light rays outside the sun is examined.) Outside this spherical hill top, the surface is concave like the surface of a saddle. This curvature is, of course, negative, but, unlike the Lobachevski geometry, it is not constant. Farther away from the center of the hill, the parabola becomes more and more similar to a straight line. The curvature is noticeably different from zero only at positions not far from the spherical portion of the surface. This negatively curved part of the surface corresponds to the part of the plane outside the sun. In the immediate vicinity of the sun, its negative curvature differs most from zero. Farther and farther away from the sun, it approaches zero. It never reaches zero, but, at a point far enough away, it is practically zero. In the diagram, the amount of curvature is greatly exaggerated. If the scale of the figure were more accurate, the curve would be so close to a straight line that curvature would not be detectable. Later, the quantitative amount will be given.

Theories T_1 and T_2, the non-Euclidean and Euclidean, may now be compared as they apply to the structure of the plane passing through the sun's center. This will be done as Helmholtz did it—by using the curved, hill-like surface as the model. Before, this was spoken of as a

Euclidean surface, which it is; but now it is being used as a model of the non-Euclidean plane. Its profile is drawn as S_1 in Figure 16–2. Below

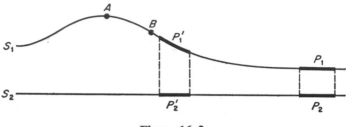

Figure 16–2.

this, the straight line S_2 represents the familiar Euclidean plane. As before, all points on S_1 are projected by parallel lines (shown as broken lines) from S_1 to S_2. Note that, if a rod moves from position P_1 to P_1', that is, from a position far from the sun to a position quite close to it, the rod does not contract, because the event is being described in the language of non-Euclidean geometry. But, if the Euclidean language of theory T_2, based on the plane S_2, is used, it must be said that the rod contracts as it moves from P_2 to P_2'. New laws must be added stating that all rods, when they are brought near the sun, suffer certain contractions in the radial direction, the direction toward the sun's center.

Figure 16–3 shows the situation as seen from above instead of in

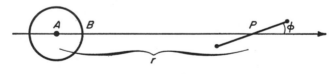

Figure 16–3.

cross section. The circle with the center at A is the sun. The rod is at position P. Let ϕ be the angle between the rod and the radial direction. The contraction of the rod, in terms of theory T_2, depends on this angle and can be covered by a general law. This law states that if a rod, which has length l_0 when it is far removed from any gravitational field, is brought (temperature and other conditions remaining unchanged) to a position P at the distance r from the body b, whose mass is m, with an angle ϕ to the radial direction, it will contract to the length

$$l_0 \left[1 - C \left(\frac{m}{r} \cos^2\phi \right) \right],$$

where C is a certain constant. Since this is a general law, as is the law of thermal expansion, it must be taken into consideration when a measuring rod that is to be used as a standard of length is defined. Therefore, a new correction term must be inserted into the equation previously used to define the length l. The definition will then be:

$$l = l_0 \left[1 + \beta(T-T_0)\right]\left[1 - C\left(\frac{m}{r}\cos^2\phi\right)\right].$$

Keep the distance r constant, but vary the angle ϕ. If the rod is in a radial direction so that $\phi = 0$, then the cosine is 1 and "$\cos^2\phi$" can be omitted from the equation. In that case, the contraction has reached its maximum value. If ϕ is a right angle, the cosine is zero, and the entire correction term disappears. In other words, there is no contraction of the rod when it is perpendicular to the radial direction. In other positions, the amount of contraction varies between zero and the maximum.

The value of the constant C is very small. If all the magnitudes are measured in the *CGS* (centimeter, gram, second) system, then the value of C is 3.7×10^{-29}. This means that behind the decimal point there are 28 zeroes followed by "37". It is apparent, then, that this is an extremely small value. Even if there is a mass as large as the sun (1.98×10^{33} grams) and if r is made as small as possible by going close to the sun's surface so that r is equal to the radius AB of the sun (6.95×10^{10} centimeters), the effect is still very small. In fact, the relative contraction of a rod near the surface of the sun, in radial direction, is

$$C\frac{m}{r_0} = .0000011.$$

It is evident, then, that the graphs of figures 16–1 and 16–2 are enormously exaggerated. The structure of a plane through the center of the sun is practically the same as that of a Euclidean plane; but there are minute deviations and, as will be shown later, there are experimental procedures for observing those deviations.

The important point to grasp here—and it is the point emphasized by Poincaré—is that the behavior of rods in gravitational fields can be described in two essentially different ways. Euclidean geometry can be preserved if we introduce new physical laws, or the rigidity of bodies can be preserved if we adopt a non-Euclidean geometry. We are free to choose whatever geometry we wish for physical space provided we are willing to make whatever adjustments are necessary in physical laws.

This adjustment applies not only to laws concerning physical bodies but to optical laws as well.

The application to optical laws can be understood easily by considering the path of a light ray that passes close to the sun as it travels from a distant star to the earth. Figure 16–4 shows the earth on the left

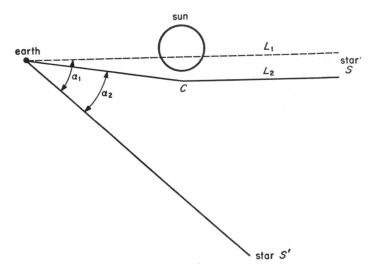

Figure 16–4.

and the sun's disk in the center. When the sun is not in the position shown, light coming from star S (the star is far outside the page to the right) would normally reach the earth along the straight line L_1. But, when the sun is in the position shown, light from the star is deflected at C, so that it takes the path L_2. Star S is so far away that the light paths L_1 and L_2 (the part to the right of point C) can be regarded as parallel. But, if an astronomer were to measure the angle a_2 between star S and another star, S', he would find it a trifle smaller than angle a_1, which he found in other seasons when the sun did not appear near star S. Thus the position of star S, as seen from the earth, appears to have shifted slightly toward star S'. This, of course, is an empirical observation, which is actually one of the basic empirical confirmations of Einstein's theory.

The sun's light is so strong that stars near its rim can be seen or photographed only during a solar eclipse. A portion of such a photograph looks something like the drawing in Figure 16–5. The position of

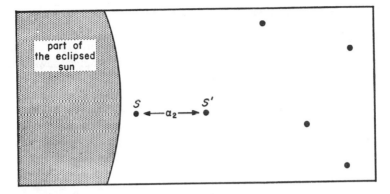

Figure 16–5.

star S is indicated by a dot. Other stars, including star S', are shown by other dots. The angle between rays of light coming from S and S' is determined by measuring the distance between S and S' on the photographic plate. This distance is then compared with the distance between the two stars on photographs taken at other times, when the sun was at some other position. Historic tests of this sort, made first in 1919 and repeated at many later eclipses, indicated a very slight shift in the positions of stars close to the sun's disk. The displacements confirmed Einstein's prediction that light rays passing near the sun would be "bent" by the sun's powerful gravitational field.

The first measurements of these displacements were made by Finlay Freundlich, in the Einstein Tower in Potsdam, near Berlin. At that time, I was living in Vienna, and I remember visiting Hans Reichenbach in Berlin; we both went to see Freundlich in the basement of the tower, where he was working. He spent many days making careful measurements of all the positions of the stars on a photographic plate about ten inches square. With the aid of a microscope, he would make repeated measurements of the coordinates of each star and would then take the mean of those measurements in order to obtain the most accurate possible estimate of the star's position. He refused to permit any of his assistants to make these measurements; he did them himself because he realized the great historic importance of the test. It turned out that the shift, although very small, could be detected, and the test proved to be a dramatic confirmation of Einstein's theory.

The situation with respect to the deflection of light rays by a gravitational field is similar to the situation with respect to the apparent contraction of physical bodies. Here again, we have to choose between two theories to explain the empirical results. In theory T_2 we keep Euclidean geometry; but then we have to devise new optical laws that will describe the deflection of light in gravitational fields. On the other hand, in theory T_1 we adopt a non-Euclidean geometry and preserve the classical assumption that, in empty space, light is not deflected by gravitational fields. This will be explained in the next chapter.

It is important to understand the nature of this choice thoroughly before asking what the geometrical structure of space is. I believe that the ambiguity of this question and the elliptical phrasing of various answers by Poincaré and others led to some misinterpretations of their position (by Reichenbach, for instance). Poincaré said that the physicist can freely choose between a Euclidean geometry and any form of non-Euclidean geometry. Because Poincaré said the choice was a matter of convention, his view became known as the conventionalist view. In my opinion, Poincaré meant that the choice was made by the physicist *before* he decided which method to use for measuring length. After making the choice, he would then *adjust* his method of measurement so that it would lead to the type of geometry he had chosen. Once a method of measurement is accepted, the question of the structure of space becomes an empirical question, to be settled by observations. Although Poincaré was not always explicit about this, his writings, taken in their entire context, indicate that this is what he meant. In my opinion, there is no difference between Reichenbach and Poincaré on this question. It is true that Reichenbach criticized Poincaré for being a conventionalist who did not see the empirical aspect of the question about the geometrical structure of space, but Poincaré was speaking elliptically; he was dealing only with the physicist's initial choice of a geometry. Both men saw clearly that once an appropriate method of measurement is adopted, the question of the geometrical structure of space becomes an empirical problem, to be answered by making observations.

The empirical aspect of this problem is brought out clearly by an interesting question that is seldom asked today but was much discussed in the early years of relativity theory. Is the total space of the universe finite or infinite? As mentioned earlier, Einstein once proposed a model of the cosmos that can be thought of as analogous to the surface of a sphere. For two-dimensional creatures on a sphere, the surface would be both finite and unbounded. It would be finite because the entire surface could be explored, and its area could be computed; it would be un-bounded in the sense that one could always move in any direction, from any position, and never encounter a boundary of any sort. In Einstein's model, three-dimensional space, viewed from a four-dimensional stand-point, would possess an overall positive curvature, so that it would close on itself like the closed surface of a sphere. A spaceship traveling in any direction in a "straight line" would eventually return to its starting point, just as an airplane, moving along a great circle of the earth, would return to its starting point. There was even speculation that a galaxy could be seen if a powerful telescope were pointed in the direction opposite to that of the galaxy.

How could Einstein think of the entire cosmos as having a positive curvature when he also maintained that in gravitational fields there was always a negative curvature? This question is still a good brain teaser to spring on a physicist. The answer is not difficult; but the question may be a puzzling one if not much thought has been given to such matters. Consider the surface of the earth. It has an overall positive curvature. Nevertheless, it is filled with valleys that have strong negative curvatures. In the same way, Einstein's cosmic model contains "valleys" of negative curvature in strong gravitational fields, but these are over-balanced by stronger positive curvatures *within* large masses, such as fixed stars. These stars correspond, in analogy with the earth's surface, to the strong positive curvatures of mountain domes. It has been calcu-lated that the cosmos could have an overall positive curvature only if its average mass density were high enough. Today, the expanding universe hypothesis and recent calculations about the amount of matter in the universe have made Einstein's closed finite model seem unlikely. Per-haps it is still an open question, because there is a great deal of uncer-tainty about measurements of masses and distances; it is possible that hydrogen may be spread throughout what was previously thought to be empty space; this would raise the average mass density of the cosmos. In any case, Einstein's attractive dream of a closed but unbounded uni-verse certainly seems less probable now than it did at the time he first proposed it. The point to be emphasized here is that the evidence for or against this cosmic model is empirical evidence. At present, although there is general acceptance of the non-Euclidean geometry of relativity theory, there is no cosmic model on which all astronomers and physicists agree.

As we have seen, physicists could have kept Euclidean geometry (as Poincaré wrongly predicted they would) and could have explained the new observations by introducing new correction factors into mechanical and optical laws. Instead, they chose to follow Einstein in his abandonment of Euclidean geometry. On what basis was this decision made? Was it for reasons of simplicity? If so, for simplicity of what? The Euclidean approach has a much simpler geometry but more complicated physical laws. The non-Euclidean approach has a vastly more complicated geometry but greatly simplified physical laws. How should a decision be made between the two approaches, each of which is simpler than the other in some respect? In the next chapter an attempt will be made to answer this question.

CHAPTER 17

Advantages of
Non-Euclidean
Physical Geometry

IN SEEKING a basis on which to choose
between a Euclidean and a non-Euclidean geometrical structure for
physical space, there is a temptation at first to choose the approach that
provides the simplest method for measuring length. In other words,
avoid, as much as is possible, the introduction of correction factors into
the methods of measurement. Unfortunately, if this rule is taken literally, the consequences are fantastic. The simplest way to measure length
is to choose a measuring rod and to define the unit of length as the
length of that rod, without introducing any correction factors at all. The
rod, regardless of its temperature, regardless of whether it is magnetized
or is acted upon by elastic forces, and regardless of whether it is in a
strong or weak gravitational field, is taken as the unit of length. As
shown earlier, there is no logical contradiction in adopting such a length
unit; nor is there any way in which this choice can be ruled out by
observed facts. However, a high price must be paid for such a choice; it
leads to a bizarre, incredibly complicated picture of the world. It would
be necessary to say, for example, that, whenever a flame is put to the
rod, all other objects in the cosmos, including the most distant galaxies,

immediately contract. No physicist would want to accept the strange consequences and involved physical laws that would result if this simplest possible definition of length were adopted.

On what basis, then, did Einstein and his followers choose the more complicated, non-Euclidean geometry? The answer is that they did not make the choice with respect to the simplicity of this or that partial aspect of the situation, but rather with respect to the overall simplicity of the total system of physics that would result from the choice. From this total point of view, we must certainly agree with Einstein that there is a gain in simplicity if non-Euclidean geometry is adopted. To preserve Euclidean geometry, physics would have to devise weird laws about the contraction and expansion of solid bodies and the deflection of light rays in gravitational fields. Once the non-Euclidean approach was adopted, there would be an enormous simplification of physical laws. In the first place, it would no longer be necessary to introduce new laws for the contraction of rigid bodies and the deflection of light rays. More than this, old laws governing the movements of physical bodies, such as the paths of planets around the sun, would be greatly simplified. Even gravitational force itself would, in a sense, disappear from the picture. Instead of a "force", there would be only the movement of an object along its natural "world-line", in a manner required by the non-Euclidean geometry of the space-time system.

The concept of the world-line can be explained in this way. Suppose that you wish to diagram on a map, *M*, the movement of your car as you drove it through the streets of Los Angeles. Figure 17–1 shows

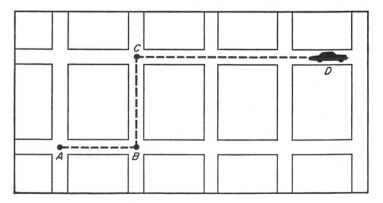

Figure 17–1.

such a map; the path of the car is indicated by the line *ABCD*. The line shows exactly how your car traveled along the streets, but, of course, it shows nothing about the speed of the car. The time element is missing.

How can the car's movement be diagrammed so that time and the velocity of the car are taken into account? This can be done by taking a series of maps, M_1, M_2, . . . , each drawn on a transparent sheet of plastic, as shown in Figure 17–2. On M_1 you mark the point A_1 (corre-

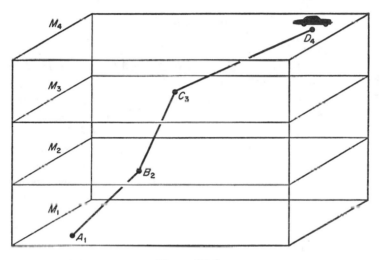

Figure 17–2.

sponding to *A* on the original map *M*), where your car was at the first time point, T_1. On M_2 you mark the car's position B_2 at a later time point, T_2 (say 20 seconds after T_1). M_3 and M_4 show the positions C_3 and D_4 of the car at time points T_3 and T_4. The maps are placed in a framework that holds them parallel, one above the other, at distances of, say, ten inches; a vertical scale of one inch for every two seconds of time is used. If a wire is placed to connect the four points, the wire will represent the *world-line* of the car's movement. In addition to showing where the car was at each moment, it will show the speed of the car as it moves from point to point.

An even simpler example of a world-line is evident when the one-dimensional path of a car driven straight along Wilshire Boulevard is shown. A world-line of this case might be drawn as shown in Figure 17–3, where the horizontal axis shows distance, and the vertical axis is

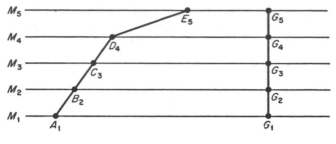

Figure 17–3.

time in minutes. The car starts at time M_1 at position A_1. For the first three minutes, the car moves at a constant speed from A_1 to D_4. From D_4 to E_5 the car's speed is constant, but is greater than before because a greater distance is covered in one minute. At the right of this chart, the world-line of a man who stood in one spot, G, during the same four minutes is shown. Since he did not move, his world-line is straight up. It is apparent that a world-line on this chart deviates more and more from the vertical as speed increases. If the speed is not constant, then the world-line is curved instead of straight. In this way, the line indicates all the features of the actual movement; even if the speed of the object is increasing or decreasing, the world-line shows its speed at every moment of time.

The world-line of an object can be diagrammed on a plane only if the object is moving along a one-dimensional path. If the path is two-dimensional, as in the first example, the world-line must be diagrammed on a three-dimensional chart. In similar fashion, the world-line of an object moving in three-dimensional space must be shown on a series of three-dimensional maps that form a four-dimensional system in the same way that the series of two-dimensional plastic maps formed a three-dimensional system. An actual model of a four-dimensional chart containing a four-dimensional world-line cannot be constructed, but the world-line can be described mathematically. A special metric introduced by Hermann Minkowski leads to an unusually simple formula. When this is applied to the laws of light rays and moving bodies, such as planets, the world-lines of both planets and light rays, in all gravitational fields, turn out to be geodesics. As explained earlier, a geodesic is the "straightest" possible line in a given space system. The space system need not have a constant curvature. On the surface of the earth, for example, with its irregular mountains and valleys, it is always possible to find one or more geodesics that represent the shortest possible paths between any two given points. Geodesics are the counterparts of straight lines on the Euclidean plane.

In relativity theory, the world-lines of planets and light rays are geodesics. Just as in classical physics, a body that is not acted upon by an external force is said to move by its inertia along a straight path with constant velocity and, therefore, along a straight world-line, so in relativity physics, this moving body is said to move, even in gravitational fields, along world-lines that are geodesics. No concept of "force" need enter this picture. Why does a planet revolve around the sun instead of moving off at a tangent? It is not because the sun is exerting a "force" that "pulls" the planet toward it, but because the sun's mass creates a negative curvature in the non-Euclidean structure of space-time. In the curved structure, the straightest world-line for the planet, its geodesic, turns out to be the one that corresponds to its actual movement around the sun. The planet's elliptical path is not a geodesic in three-dimensional space, but its world-line, in the four-dimensional non-Euclidean space-time system, is a geodesic. It is the straightest possible line the planet can take. In similar fashion, light also travels through space-time along geodesic world-lines.

From the non-Euclidean view of relativity theory, there is no force of gravity in the sense of elastic or electromagnetic forces. Gravitation, as force, vanishes from physics and is replaced by the geometrical structure of a four-dimensional space-time system. This was such a revolutionary transformation that it is not hard to understand why many failed to grasp the concept correctly. It was sometimes said that a part of physics, namely, the theory of gravitation, had been replaced by pure geometry, or that part of physics had turned into mathematics. Some writers speculated on the possibility that some day the whole of physics might turn into mathematics. I think this is misleading. Writers who try to make relativity theory clearer to the layman enjoy using stimulating, paradoxical phrases. Such phrases may contribute to colorful writing, but they often give an inaccurate impression of the true state of affairs. In this case, I think they lead to a confusion between geometry in its mathematical sense and geometry in its physical sense. The physics of gravitation is indeed replaced, in relativity theory, by a physical geometry of space or, more accurately, of the space-time system. But this geometry is still a part of physics, not of pure mathematics. It is *physical,* not mathematical geometry.

Mathematical geometry is purely logical, whereas physical geometry is an empirical theory. In Einstein's theory of relativity, gravitation simply took another form. One physical theory of gravity was transformed into another physical theory. The concept of force no longer applies, but the relativity theory of gravitation is still physics, not mathematics. Non-mathematical magnitudes (distributions of the curvature of space-time) continue to occur within it. These are physical magnitudes, not mathematical concepts. The point to be emphasized here is that, because Einstein's theory of gravitation was called geometry, there was a temptation to view it as if it were pure mathematics. But physical geometry is not mathematics; it is a theory of physical space. It is not just an empty abstraction. It is the physical theory of the behavior of bodies and light rays and therefore cannot possibly be regarded as part of pure mathematics. It has been mentioned before that one should take *cum grano salis* Galileo's famous remark that the book of nature is written in the language of mathematics. This remark is easily misunderstood. Galileo meant that nature can be described with the help of mathematical concepts, not that the total language of physics consists of mathematical symbols. It is absolutely impossible to define a concept such as "mass" or "temperature" in pure mathematics in the way that the concept of logarithm or any other mathematical function can be defined. It is essential to realize that there is a fundamental difference between the physical symbols occurring in a physical law (for example, *"m"* for mass, *"T"* for temperature) and the mathematical symbols that occur in the law (for example, "2", "$\sqrt{\ }$", "log", "cos").

The great simplicity of Einstein's equations for moving bodies and light rays is certainly in favor of his claim that the non-Euclidean approach is preferable to the Euclidean one, in which it would be necessary to complicate the equations by introducing new correction factors. But this is still far from the discovery of any sort of general principle that will tell how to obtain the greatest overall simplicity in choosing between alternative approaches to physics. What is desired is a general rule of choice that can be applied in all future situations; Einstein's choice in this situation would then be a special case of the general rule. It is taken for granted, of course, that the simplest overall system of physics is preferable, but that is not the question. The question is how to decide which of two systems has the maximum overall simplicity. When there are two competing systems, it is often the case that each is in some respect simpler than the other. In such cases, how can the overall simplicity be measured?

It was Reichenbach's merit to have proposed a general rule of this sort. Perhaps his rule is not quite absolutely general, but it covers a comprehensive class of situations and is very interesting. I have the impression that insufficient attention has been paid to it. The rule is based on a distinction between "differential forces" and "universal forces". Reichenbach called them "forces", but it is preferable here to speak of them, in a more general way, as two kinds of "effects". (Forces can be introduced later to explain the effects.) The distinction is this. If an effect is different with respect to different substances, it is a *differential effect*. If it is quantitatively the same, regardless of the nature of the substance, it is a *universal effect*.

This can be made clear by examples. When an iron rod is heated, it expands. If length is defined by means of an iron rod, this effect of thermal expansion is taken into account (as shown earlier) by the introduction of a correction factor:

$$l = l_0 [1 + \beta(T - T_0)].$$

The beta in this formula is the coefficient of thermal expansion. It is a constant, but only for all bodies of a certain substance. If the rod is iron, beta has a certain value; if it is copper, gold, or some other substance, it has different values. The expansion of the rod when heated is, therefore, clearly a differential effect, because it varies with the substance.

Consider the formula for length after a second correction factor has been added; this one takes into account the influence of gravitation on the length of the rod. The formula, it may be recalled, is:

$$l = l_0 [1 + \beta(T - T_0)] [1 - C(\frac{m}{r} \cos^2\phi)].$$

The C in this second correction factor is a universal constant, which is the same in every gravitational field and with respect to any body. There is no parameter inside the right-hand pair of brackets that changes from substance to substance in the way that the parameter beta, inside the first pair of brackets, changes. The correction factor takes into consideration the mass m of the sun, the distance r from the sun to the measuring rod, and the angle ϕ of the rod with respect to a radial line from sun to rod. It indicates nothing whatever about whether the rod is iron, copper, or some other substance. It is, therefore, a universal effect.

Reichenbach sometimes added that the universal effects are of such a kind that shields cannot be made against them. A metal rod, for example, can be shielded from magnetic effects by surrounding it with a wall of iron. But there is no way to shield it from gravitational effects. In my opinion, it is not necessary to speak of shields in order to distinguish between differential and universal effects, because this condition is already implied in what has been said before. If an iron wall is built to shield a piece of apparatus from a strong magnet in the next room, the shield is effective only because the iron wall is influenced by magnetic fields differently than the air is influenced. If this were not so, the shield would not work. The concept of shielding applies, therefore, only to effects that have different influences on different substances. If a universal effect is defined as one that is the same for all substances, it follows that no shielding from the effect is possible.

In a detailed analysis of differential and universal effects,[1] Reichenbach calls special attention to the following fact. Suppose someone states that he has just discovered a new effect and says that it does not vary from substance to substance. The law he gives for this new effect is examined, and it is apparent that what he says is true; the law contains no parameter that varies with the nature of the substance. In cases of this sort, Reichenbach maintained, the theory can always be reformulated so that the universal effect will completely disappear.

There is no comparable way to eliminate a differential effect, such as thermal expansion. The assertion that there are no thermal expansion effects can easily be disproved. Simply place two rods of different substances alongside each other, heat them to the same higher temperature, and observe the resulting difference in lengths. Clearly, something has changed, and there is no way to account for this observed difference without introducing the concept of thermal expansion. On the other hand, a universal effect such as the influence of gravity on the lengths of rods *can* be accounted for by adopting a theory in which the effect vanishes entirely. This is exactly what happens in Einstein's theory of relativity. Adoption of a suitable non-Euclidean space-time system removes the need to speak of bodies expanding and contracting in gravitational fields. Bodies do not alter their sizes when moved around in such fields; but in this theory there is a different structure of space-time. Unlike the previous situation with respect to thermal expansion, there is no way to show that the elimination of this gravitational effect is impossible. Gravitational fields have exactly the same effect on all substances. If two rods

[1] See Section 6, "The Distinction between Universal and Differential Forces," in Hans Reichenbach, *The Philosophy of Space and Time* (New York: Dover, 1958).

are placed alongside each other and turned in various directions, they remain exactly the same length with respect to each other.

In view of these considerations, Reichenbach proposed this rule for simplifying physical theory: Whenever there is a system of physics, in which a certain universal effect is asserted by a law that specifies under what condition and in what amount the effect occurs, the theory should be transformed so that the amount of the effect will be reduced to zero. This is what Einstein did in regard to the contraction and expansion of bodies in gravitational fields. From the Euclidean point of view, such changes do occur, but they are found to be universal effects. However, adoption of the non-Euclidean space-time system causes these effects to become zero. Certain other effects, such as that the angles of a triangle no longer sum to 180 degrees, may be found, but it is no longer necessary to speak of expansions and contractions of rigid bodies. Whenever universal effects are found in physics, Reichenbach maintained, it is always possible to eliminate them by a suitable transformation of theory; such a transformation should be made because of the gain in overall simplicity that would result. This is a useful general principle, deserving more attention than it has received. It applies not only to relativity theory, but also to situations that may arise in the future in which other universal effects may be discovered. Without the adoption of this rule, there is no way to give a unique answer to the question, What is the structure of space? If the rule is adopted, this question is no longer ambiguous.

When Einstein first proposed a non-Euclidean geometry for space, strong objections were raised. The objection of Dingler and others, that Euclidean geometry was indispensable because it was already presupposed in the construction of measuring instruments, has already been mentioned, but, as has been shown, that objection is certainly wrong. A more common objection, from a more philosophical point of view, was that non-Euclidean geometry should not be adopted, because it is impossible to imagine it. It is contrary to our ways of thinking, to our intuition. This objection was sometimes expressed in a Kantian way, sometimes in a phenomenological way (the terminology differed), but in general the point was that our minds seem to work in such fashion that we cannot visualize any sort of non-Euclidean spatial structure.

This point is also discussed by Reichenbach.[2] I think he is right in calling it a psychological problem and in saying that there are no

[2] *Ibid.*, Sections 9–11.

grounds for assuming that our intuitions have been preshaped in a Euclidean way. There are, on the contrary, excellent reasons for believing that the visual space, at least the visual space of a child, is non-Euclidean. "Spatial intuition", as it is called, is not so much an intuition of a metric structure as an intuition of a topological structure. Our perceptions tell us that space is three-dimensional and continuous and that every point has the same topological properties as any other point. But, with regard to the metric properties of space, our intuitions are vague and inexact guides.

The non-Euclidean character of space perception is indicated by the mind's surprising ability to adjust to whatever type of images appear on the retina. A person with strong astigmatism, for example, will have strongly distorted images on the retina of each eye. His retinal images of a yardstick may be longer when he views a horizontally placed stick than when he views the same stick placed vertically, but he is unaware of this, because the lengths of all objects in his visual field are altered in a similar way. When this person is first fitted with corrective glasses, his visual field will appear distorted for many days or weeks until his brain has adjusted to the normal images on his retina. Similarly, a person with normal vision can wear special glasses that distort images along one coordinate; after a time he becomes accustomed to the new images, and his visual field appears normal. Helmholtz described experiments of this sort, some of which he actually carried out, from which he concluded that visual space can have a non-Euclidean structure. Helmholtz believed—and I think good arguments can be made for this belief—that if a child or even an adult were sufficiently conditioned to experiences involving the behavior of bodies in a non-Euclidean world, he would be able to visualize non-Euclidean structure with the same ease that he now can visualize Euclidean structure.

Even if this belief of Helmholtz's is unfounded, there is a more essential argument against the objection that non-Euclidean geometry should not be adopted because it cannot be imagined. The ability to visualize is a psychological matter, entirely irrelevant to physics. The construction of physical theory is not limited by man's power to visualize; in fact, modern physics has moved steadily away from what can be directly observed and imagined. Even if relativity theory contained much stronger deviations from intuition and it turned out that our spatial intuition has a permanent and unchangeable Euclidean bias, we could still use in physics whatever geometrical structure we desire.

In the nineteenth century, in England more than on the continent, there was a strong effort in physics toward visualization and the construction of models. The ether was represented as a strange kind of transparent, jelly-like substance capable of oscillating and transmitting electromagnetic waves. As physics advanced, this model of the ether became more and more complicated and even acquired properties that seemed incompatible. For example, the ether had to be thought of as completely without density, because it offered no observable resistance to the motions of planets and satellites; yet light waves were found to be transverse rather than longitudinal, more like what would be expected in bodies of extremely *high* density. Although these properties were not logically incompatible, they made it very difficult to develop an intuitively satisfying model of the ether. Eventually, the various ether models became so complex that they no longer served any useful purpose. This is why Einstein found it best to abandon the ether entirely. It was simpler to accept the equations—the Maxwell and Lorentz equations—and to calculate with them instead of trying to build a model so bizarre that it was of no help in visualizing the structure of space.

Not only the ether was given up. The nineteenth-century tendency to construct visual models became weaker and weaker as twentieth-century physics advanced. The newer theories were so abstract that they had to be accepted entirely in their own terms. The psi-functions, representing the states of a physical system, such as an atom, are too complicated to permit models that can be easily visualized. Of course, it is often possible for a skillful teacher or writer on scientific topics to use a diagram that is helpful in explaining some aspect of an abstruse theory. There is no objection to the use of such diagrams as teaching devices. The point to be stressed is that it is not a valid objection to a new physical theory to say that it is more difficult to visualize than an old one. This is exactly the sort of objection that was often raised against relativity theory when it was first proposed. I remember an occasion, about 1930, when I discussed relativity with a German physicist in Prague. He was extremely depressed.

"This is terrible", he said. "Look at what Einstein has done to our wonderful physics!"

"Terrible?" I replied. I was enthusiastic about the new physics. With only a few general principles describing a certain type of invariance and the exciting adoption of non-Euclidean geometry, so much could be explained that had been unintelligible before! But this physicist had so strong an emotional resistance to theories difficult to visualize that he had almost lost his enthusiasm for physics because of Einstein's revolutionary changes. The only thing that sustained him was the hope that some day—and he hoped it would be during his lifetime—a counter-revolutionary leader would come to restore the old classical order, in which he could breathe comfortably and feel at home again.

A similar revolution took place in atomic physics. For many years, it was pleasant and satisfying to have Niels Bohr's model of the atom; a kind of planetary system with a nucleus in the center and the electrons moving around it in orbits. But it proved to be an oversimplification. The nuclear physicist today does not even try to make a total model. If he uses a model at all, he is always aware that it pictures only certain aspects of the situation and leaves out other aspects. The total system of physics is no longer required to be such that all parts of its structure can be clearly visualized. This is the fundamental reason why the psychological statement that it is not possible to visualize non-Euclidean geometry, even if true (and in my opinion it is doubtful), is not a valid objection to the adoption of a non-Euclidean physical system.

A physicist must always guard against taking a visual model as more than a pedagogical device or makeshift help. At the same time, he must also be alert to the possibility that a visual model can, and sometimes does, turn out to be literally accurate. Nature sometimes springs such surprises. Many years before physics developed any clear notions about how atoms were linked together in molecules, it was a common practice to draw schematic pictures of molecular structure. The atoms of a substance were indicated by capital letters, and valence lines were drawn to connect them in various ways. I recall talking to a chemist who objected at the time to such diagrams.

"But are they not a great help?" I asked.

"Yes", he said, "but we must warn our students not to think of these diagrams as representing actual spatial configurations. We really do not know anything at all about spatial structure on the molecular level. These diagrams are no more than diagrams, like a curve on a graph to illustrate an increase in population or pig-iron production. We all know that such a curve is only a metaphor. The population or the pig-iron is not rising in any spatial sense. Molecular pictures must be thought of in the same way. No one knows what sort of actual spatial structure molecules have."

I agreed with the chemist, but I argued that there was the possibility, at least, that molecules might be linked together in just the way the diagrams indicated, especially in view of the fact that stereoisomers had been discovered, which made it convenient to think of one molecule as a mirror image of another. If one kind of sugar twists polarized light clockwise, and another type of sugar twists it counterclockwise, then some sort of spatial configuration of atoms in the molecules seems to be indicated; configurations capable of having right- and left-handed forms.

"It is true", he replied, "that this is suggested. But we do not know for sure that this is the case."

He was right. At that time, so little was known about molecular structure that it would have been premature to insist that, as more and more was learned about such structure, it would continue to be possible to represent molecules by visualizable three-dimensional models. It was conceivable that later observations would require structures of four, five, or six dimensions. The diagrams were no more than convenient pictures of what was then known.

But it soon turned out, particularly after Max von Laue's determination of crystal structures by means of X-ray diffraction, that the atoms in molecular compounds actually *are* spatially situated in the way shown by the structure diagram. Today, a chemist does not hesitate to say that, in a protein molecule, there are certain atoms here and certain other atoms there and that they are arranged in the form of a helix. Models showing the linkages of atoms in three-dimensional space are taken quite literally. No evidence to dispute this has been found, and there are excel lent reasons to think that three-dimensional models of molecules represent actual configurations in three-dimensional space. Something of the same surprise occurred more recently as a consequence of experiments showing that parity is not conserved in the weak nuclear interactions. It now appears that particles and antiparticles, hitherto regarded as mirror images only in a metaphorical sense, may actually be mirror images in a spatial sense.

Therefore, the warning against taking models literally, although correct in principle, may later prove unnecessary. A theory may move away from models that can be visualized; then, in a later phase, when more is known, it may move back again to visual models that were previously doubted. In the case of molecular models, it was chiefly the physicists who doubted. The picture of atoms spatially arranged in molecules is so convenient that most chemists interpreted the models literally, although the physicists were correctly saying that there was not yet sufficient justification for it.

Models in the sense of visual spatial structures should not be confused with models in the modern mathematical sense. Today, the common practice of mathematicians, logicians, and scientists is to speak of models when they mean an abstract conceptual structure, not something that can be built in the laboratory with balls and wires. This model may be only a mathematical equation or set of equations. It is a simplified description of any structure—physical, economic, sociological, or other —in which abstract concepts can be related in a mathematical way. It is a simplified description because it leaves out many factors that would otherwise complicate the model. The economist, for example, speaks of one model for free market economics, another for planned economics, and so on. The psychologist speaks of a mathematical model of the learning process, of how one psychological state is related to another, with certain transitional probabilities that make the series one that mathematicians call a Markov chain. These are entirely different from the models of nineteenth-century physics. The purpose in making them is not to visualize but to formalize. The model is purely hypothetical. Certain parameters are put into it and adjusted until the best fit with the data is obtained. As more observations are made, it may turn out that the parameters not only have to be adjusted further, but also that the basic equations need to be changed. In other words, the model itself is altered. The old model served fairly well for a time; now a new model is called for.

The nineteenth-century physical model was not a model in this abstract sense. It was intended to be a spatial model of a spatial structure, in the same way that a model ship or airplane represents an actual ship or plane. Of course, the chemist does not think that molecules are made up of little colored balls held together by wires; there are many features of his model that are not to be taken literally. But, in its general spatial configuration, it is regarded as a correct picture of the spatial configuration of the atoms of the actual molecule. As has been shown, there are good reasons sometimes for taking such a model literally—a model of the solar system, for example, or of a crystal or molecule. Even when there are no grounds for such an interpretation, visual models can be extremely useful. The mind works intuitively, and it is often helpful for a scientist to think with the aid of visual pictures. At the same time, there must always be an awareness of a model's limitations. The building of a neat visual model is no guarantee of a theory's soundness, nor is the lack of a visual model an adequate reason to reject a theory.

CHAPTER **18**

Kant's Synthetic A Priori

IS IT POSSIBLE for knowledge to be both synthetic and a priori? This famous question was asked by Immanuel Kant and answered by him in the affirmative. It is important to understand exactly what Kant meant by his question and why contemporary empiricists disagree with his answer.

Two important distinctions are involved in Kant's question: a distinction between *analytic* and *synthetic* and one between *a priori* and *a posteriori*. Various interpretations have been made of both distinctions. In my opinion, the first is logical and the second is epistemological.

First, consider the logical distinction. Logic is concerned solely with whether a statement is true or false on the basis of meanings assigned to the statement's terms. For example, define the term "dog" as follows: *"X is a dog if and only if X is an animal having certain characteristics."* To be an animal, therefore, is part of the meaning of the term "dog". If, on the basis of this understanding, the assertion is made that "All dogs are animals", this would be what Kant called an analytic judgment. It involves nothing more than the meaning relations of the terms. Kant did not put it quite this way, but this is essentially what he meant. On the other hand, a synthetic statement, such as, "The moon revolves around the earth", has a factual content. As with most scientific statements, it is synthetic because it goes beyond the assigned meanings of the terms. It tells something about the nature of the world.

The distinction between a priori and a posteriori is an epistemological distinction between two kinds of knowledge. By a priori Kant meant the kind of knowledge that is independent of experience, but not independent in a genetic or psychological sense. He was fully aware that all human knowledge depends in a genetic sense on experience. Without experience, there would obviously be no knowledge of any sort. But certain kinds of knowledge are supported by experience in a way that is not true of other kinds. Consider, for example, the analytic statement, "All dogs are animals." It is not necessary to observe dogs in order to make this assertion; indeed, it is not even necessary for dogs to exist. It is only necessary to be able to conceive of a thing such as a dog, which has been defined in a way that makes being an animal a part of the definition. All analytic statements are a priori in this sense. It is not necessary to refer to experience in order to justify them. True, it may be that our experience with dogs has led us to conclude that dogs are animals. In a wide sense of the word experience, everything that we know is based on experience. The important point is that it is never necessary to refer to experience as a justification for the truth of an analytic statement. It need not be said that, "Yesterday I examined some dogs and some nondogs; then I examined some animals and some nonanimals; and I finally concluded, on the basis of this investigation, that all dogs are animals." On the contrary, the statement, "All dogs are animals", is justified by pointing out that in our language, the term "dog" is understood to have a meaning that includes "being an animal". It is justified in the same way that the analytic truth of the statement, "A unicorn has a single horn on his head", is justified. The meanings of the terms imply the truth of the statement, without reference to any examination of the world.

In contrast, a posteriori statements are assertions that cannot be justified without reference to experience. Consider, for example, the statement that the moon revolves around the earth. Its truth cannot be justified by citing the meaning of such terms as "moon", "earth", and "revolves around". Literally, of course, "a priori" and "a posteriori" mean "from prior" and "from posterior", but Kant made it perfectly clear that he did not mean this in a temporal sense. He did not mean that, in a posteriori knowledge, experience has occurred *before* the knowledge is acquired; in this sense, of course, experience is prior to *all* knowledge. He meant that experience is an essential *reason* for asserting a posteriori knowledge. Without certain specified experiences (in the case of the moon's revolution around the earth, these experiences are various astronomical observations), it is not possible to justify an a posteriori statement. In a rough sense, a posteriori knowledge today would be called empirical knowledge; it is knowledge that is essentially dependent on experience. A priori knowledge is independent of experience.

As stated earlier, all analytic statements are clearly a priori. But now an important question arises. Does the boundary line between a priori and a posteriori coincide with the boundary line between analytic and synthetic? If the two lines coincide, they can be diagrammed as shown in Figure 18–1. But perhaps the boundaries do not coincide. The

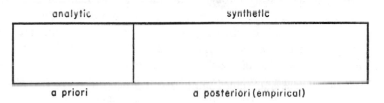

Figure 18–1.

line between a priori and a posteriori cannot lie to the left of the line between analytic and synthetic (because all analytic statements are also a priori), but it can lie to the right, as shown in Figure 18–2. If so, then

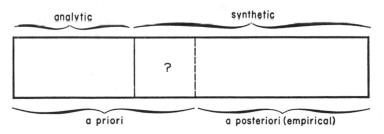

Figure 18–2.

there is an intermediate region where the synthetic overlaps the a priori. This is Kant's view. There is, he maintained, a realm of knowledge that is both synthetic and a priori. It is synthetic because it tells something about the world, and it is a priori because it can be known with certainty, in a way that does not call for justification by experience. Is there such a region? This is one of the great controversial questions in the history of the philosophy of science. Indeed, as Moritz Schlick once remarked, empiricism can be defined as the point of view that maintains that there is no synthetic a priori. If the whole of empiricism is to be compressed into a nutshell, this is one way of doing it.

Geometry provided Kant with one of his chief examples of synthetic a priori knowledge. His reasoning was that if the axioms of geometry (by which he meant Euclidean geometry—no other geometry was available in his time) are considered, it is not possible to imagine the axioms as not true. For instance, there is one and only one straight line through two points. Intuition, here, gives absolute certainty. It is possible to imagine a straight line connecting two points, but any other line conceived of as passing through them must be curved, not straight. Therefore, Kant argued, we have the right to complete confidence in the knowledge of all axioms of geometry. Since the theorems are all logically derived from the axioms, we are also entitled to complete confidence in the truth of the theorems. Geometry, therefore, is completely certain in a way that does not demand justification by experience. It is not necessary to make points on a sheet of paper and draw various lines in order to establish the statement that only one straight line will connect two points. It is justified by intuition; and, although a geometrical theorem may be very complicated and not at all obvious, it can be justified by proceeding from the axioms by logical steps that are also intuitively certain. In short, all geometry is a priori.

On the other hand, Kant continued, the theorems of geometry tell something about the world. Consider the theorem that the sum of the interior angles of a triangle is 180 degrees. This can be derived logically from Euclidean axioms, so there is a priori knowledge of its truth. But it is also true that, if a triangle is drawn and its angles measured, they are found to add up to 180 degrees. If the sum deviates from this, a more careful examination of the construction will always reveal that the lines were not perfectly straight or that, perhaps, the measurements were inaccurate. The theorems of geometry, then, are more than a priori statements. They describe the actual structure of the world and, therefore, are also synthetic. Yet, clearly they are not a posteriori in the way scientific laws are. A scientific law has to be justified by experience. It is easy to imagine that tomorrow an event could be observed that would contradict any given scientific law. It is easy to suppose that the earth might go around the moon, instead of vice versa, and it can never be certain that tomorrow science might not make discoveries that would require a modification of what was previously supposed to be true. But this is not the case with geometrical laws. It is inconceivable that new discoveries in geometry could modify the truth of the Pythagorean theorem. Euclidean geometry is intuitively certain, independent of experience. In geometry, Kant was convinced, we have a paradigm of the union of synthetic and a priori knowledge.

From a modern point of view, the situation looks quite different. Kant should not be blamed for his error because, in his day, non-Euclidean geometry had not been discovered. It was not possible for him to think about geometry in any other way. In fact, throughout the entire nineteenth century, except for a few bold individuals, such as Gauss, Riemann, and Helmholtz, even mathematicians took this Kantian point of view for granted. Today, it is easy to see the source of Kant's error. It was a failure to realize that there are two essentially different kinds of geometry—one mathematical, the other physical.

Mathematical geometry is pure mathematics. In Kantian terms, it is indeed both analytic and a priori. But it is not possible to say that it is also synthetic. It is simply a deductive system based on certain axioms that do not have to be interpreted by reference to any existing world. This can be demonstrated in many different ways, one of which is given in Bertrand Russell's early book, *The Principles of Mathematics* (not to be confused with the later *Principia Mathematica*).[1] Russell shows how it is possible to define Euclidean space entirely as a system of primitive relations for which certain structural properties are assumed; for example, one relation is symmetric and transitive, another is asymmetric, and so on. On the basis of these assumptions it is possible to derive logically a set of theorems for Euclidean space, theorems that comprise the whole of Euclidean geometry. This geometry says nothing at all about the world. It says only that, if a certain system of relations has certain structural properties, the system will have certain other characteristics that follow logically from the assumed structure. Mathematical geometry is a theory of logical structure. It is completely independent of

[1] See Part VI of *The Principles of Mathematics* (Cambridge: Cambridge University Press, 1903); (2nd ed., with new introduction, London: Allen & Unwin. 1938); (New York: Norton, 1938).

scientific investigations; concerned solely with the logical implications of a given set of axioms.

Physical geometry, on the other hand, is concerned with the application of pure geometry to the world. Here the terms of Euclidean geometry have their ordinary meaning. A point is an actual position in physical space. Of course we cannot observe a geometrical point, but we can approximate it by making, say, a tiny spot of ink on a sheet of paper. In a similar way, we can observe and work with approximations of lines, planes, cubes, and so on. These words refer to actual structures in the physical space we inhabit and are also part of the language of pure or mathematical geometry; here, then, lies a primary source of nineteenth-century confusion about geometry. Because the same words were used by the scientist and by the pure mathematician, it was wrongly assumed that both were making use of the same kind of geometry.

The distinction between the two geometries became especially clear through David Hilbert's famous work on the foundations of geometry.[2] "We are thinking here of three distinct systems of things", Hilbert wrote. "The things of the first system we will call *points,* those of the second system *lines,* and those of the third system *planes.*" Although he called these entities by the names of "points", "lines", and "planes", he implied nothing whatever about the meaning of these words. They were convenient to use only because they were familiar and provided the reader with a visualization of one possible interpretation of the terms. But the geometrical system, as Hilbert constructed it, was entirely free of any interpretation. "Points", "lines", and "planes" could be taken to mean any three classes of entities that fulfilled the relations stated in the axioms. For example, instead of physical points, lines, and planes, one could interpret "point" as an ordered triple of real numbers. A "line" would then be a class of ordered triples of real numbers that fulfilled two linear equations, and a "plane" would be a class of ordered triples that fulfilled one linear equation. In pure or mathematical geometry, terms such as "points", "lines", and "planes" are not used in the ordinary sense. They have an infinity of possible interpretations.

Once this distinction between pure and physical geometry is understood, it becomes clear how Kant's belief, and the beliefs of almost all nineteenth-century philosophers, involved a fundamental confusion be-

---

[2] Hilbert's *Grundlagen der Geometrie* ("Foundations of Geometry") first appeared in Germany in 1899. An English translation by E. J. Townsend was published in Chicago by Open Court (1902) and is currently available as a paperback.

tween two fields of quite different character. When we say, "Geometry is certainly a priori; there is no doubt about the truth of its theorems", we are thinking of mathematical geometry. But suppose we add, "It also tells us something about the world. With its help, we can predict the outcome of measurements of actual geometrical structures." Now we have inadvertently slipped over to the other meaning of geometry. We are talking about physical geometry; about the structure of actual space. Mathematical geometry is a priori. Physical geometry is synthetic. No geometry is both. Indeed, if empiricism is accepted, there is no knowledge of any sort that is both a priori and synthetic.

In reference to knowledge in geometry, the distinction between the two kinds of geometry is fundamental and is now universally recognized. When a challenge is made about the nature of geometrical knowledge, the first question to ask is: "Which type of geometry do you have in mind? Are you speaking of mathematical or physical geometry?" A clear distinction here is essential if confusion is to be avoided and if the revolutionary advances in the theory of relativity are to be understood.

One of the clearest, most precise statements of this distinction was made by Einstein at the close of a lecture entitled "Geometry and Experience." [3] Einstein spoke of "mathematics", but he meant geometry in the two ways that it can be understood. "So far as the theorems of mathematics are about reality", he said, "they are not certain." In Kantian terminology, this means that so far as they are synthetic, they are not a priori. "And so far as they are certain", he continued, "they are not about reality." In Kantian terminology, so far as they are a priori, they are not synthetic.

Kant held that a priori knowledge is certain knowledge; it cannot be contradicted by experience. Relativity theory made it clear, to all who understood, that, if geometry is taken in this a priori sense, it tells us nothing about reality. No statement is possible that combines logical certainty with knowledge of the geometrical structure of the world.

[3] Einstein's lecture was published separately as *Geometrie und Erfahrung* (Berlin: 1921), later translated and included in Albert Einstein, *Sidelights on Relativity* (New York: Dutton, 1923).

Part **IV**
CAUSALITY
AND DETERMINISM

CHAPTER 19

Causality

THE CONCEPT OF causality, one of the central topics in today's philosophy of science, has occupied the attention of distinguished philosophers from the time of the ancient Greeks down to the present. In previous periods, it was a topic in what was called the philosophy of nature. That field embraced both the empirical investigation of nature and the philosophical clarification of such knowledge. Today, it has become increasingly clear that the investigation of nature is the task of the empirical scientist, not the task of the philosopher as such.

Of course, a philosopher can be both a philosopher and a scientist. If such is the case, he should be aware of a fundamental difference between two kinds of questions that he can ask. If he asks questions such as. "How were the moon's craters formed?" or "Is there a galaxy composed of antimatter?", he is posing questions for astronomers and physicists. On the other hand, if he directs his questions, not toward the nature of the world, but toward an analysis of the fundamental concepts of a science, then he is posing questions in the philosophy of science.

In previous periods, philosophers believed that there was a metaphysics of nature, a field of knowledge deeper and more fundamental than any empirical science. The philosopher's task was to expound metaphysical truths. Today's philosophers of science do not believe there is such a metaphysics. The old philosophy of nature has been replaced by the philosophy of science. This newer philosophy is not concerned with the discovery of facts and laws (the task of the empirical scientist), nor with the formulation of a metaphysics about the world. Instead, it turns its attention toward science itself, studying the concepts employed, methods used, possible results, forms of statements, and types of logic that are applicable. In other words, it is concerned with the sort of problems discussed in this book. The philosopher of science studies the philosophical (that is, the logical and methodological) foundations of psychology, not the "nature of the mind". He studies the philosophical foundations of anthropology, not the "nature of culture". In each field, the primary concern is with the concepts and methods of that field.

Some philosophers have warned against drawing too sharp a distinction between the work of scientists in a given field and the work of a philosopher of science who concerns himself with that field. In a sense, this warning is a good one. Although the work of the empirical scientist and the work of the philosopher of science must always be distinguished, in practice the two fields usually intermingle. A working physicist is constantly coming upon methodological questions. What sort of concepts should he use? What rules govern these concepts? By what logical method can he define his concepts? How can he put his concepts together into statements and the statements into a logically connected system or theory? All these questions he must answer as a philosopher of science; clearly, they cannot be answered by empirical procedures. On the other hand, it is impossible to do significant work in the philosophy of science without knowing a great deal about the empirical results of science. In this book, for example, it has been necessary to speak at length about some particular features of relativity theory. Other details about the theory were not discussed because the theory was introduced primarily to clarify the important distinction between empirical geometry and pure or mathematical geometry. Unless a student of the philosophy of science thoroughly understands a science, he cannot even raise important questions about its concepts and methods.

My reason for distinguishing the task of the philosopher of science from the metaphysical task of his predecessor, the philosopher of nature, is that this distinction is important for the analysis of causality, the topic of this chapter. The older philosophers were concerned with the metaphysical nature of causality itself. Our concern here is to study how empirical scientists make use of the concept of causality, to make clear precisely what they mean when they say, "This is the cause of that." Exactly what does the cause-and-effect relation mean? In everyday life, the concept is certainly a vague one. Even in science, it is often not clear what a scientist means when he says that one event has "caused" another. One of the most important tasks of the philosophy of science is to analyze the concept of causality and to clarify its meaning.

Even the historical origin of the concept is somewhat vague. It apparently arose as a kind of projection of human experience into the world of nature. When a table is pushed, tension is felt in the muscles. When something similar is observed in nature, such as one billiard ball striking another, it is easy to imagine that one ball is having an experience analogous to our experience of pushing the table. The striking ball is the agent. It *does* something to the other ball that makes it move. It is easy to see how men of primitive cultures could suppose that elements in nature were animated, as they themselves were, by souls that willed certain things to happen. This is especially understandable with respect to natural phenomena that cause great harm. A mountain would be blamed for causing a landslide. A tornado would be blamed for damaging a village.

Today, this anthropomorphic approach to nature is no longer held by civilized men and certainly not by scientists. Nevertheless, elements of animistic thinking tend to persist. A stone shatters a window. Did the stone intend to do this? Of course not, the scientist will say. A stone is a stone. It possesses no soul capable of intention. On the other hand, most people, even the scientist himself, will not hesitate to say that event *b,* the breaking of the window, was *caused* by event *a,* the collision of the stone with the glass. What does the scientist mean when he says that event *b* was caused by event *a?* He might say that event *a* "brought about" event *b* or "produced" event *b.* So you see, when he tries to explain the meaning of "cause", he falls back on such phrases as "bring about", "bring forth", "create", and "produce". Those are metaphorical phrases, taken from human activity. A human activity can, in a literal sense, bring forth, create, and produce various other events; but in the case of the stone, this cannot be taken literally. It is not a very satisfactory answer to the question: "What does it mean to say that one event caused another?"

It is important to analyze this vague concept of causality, to purify it of all the old, unscientific components that may be involved. But first, one point should be made clear: I do not believe there is any reason to reject the concept of causality. Some philosophers contend that David Hume, in his famous critique of causality, meant to reject the concept *in toto*. I do not believe this was Hume's intention. He did not mean to reject the concept, but only to purify it. Later this question will be considered again, but now I want to say that what Hume rejected was the component of necessity in the concept of causality. His analysis was in the right direction, although, in the opinion of today's philosophers of science, it did not go far enough; nor was it sufficiently clear. In my opinion, it is not necessary to regard causality as a prescientific concept, metaphysical in a derogatory sense and therefore to be discarded. After the concept has been analyzed and fully explicated, it will be found that something remains that can be called causality; this something justifies its use for centuries, both by scientists and in everyday life.

We begin the analysis by asking: between what kinds of entities does the causal relation hold? Strictly speaking, it is not a *thing* that causes an event, but a process. In everyday life we speak of certain things causing events. What we really mean is that certain processes or events cause other processes or events. We say the sun causes plants to grow. What we really mean is that radiation from the sun, a process, is the cause. But if we make "processes" or "events" the entities involved in cause-and-effect relations, we must define these terms in an extremely wide sense. We must include, as we do *not* in everyday life,· processes that are static.

Consider, for example, a table. I can observe nothing about it that is changing. Yesterday it may have been moved, in the future it may be damaged or destroyed, but at the moment I observe no change. It can be assumed that its temperature, mass, even the reflection of light on its surface, and so on remain unchanged for a certain period. This event, the table existing without change, is also a process. It is a *static* process, one in which the relevant magnitudes remain constant in time. If processes or events are spoken of as involved in cause-effect relations, it must be recognized that these terms include static processes; they stand for any sequence of states of a physical system, both changing and unchanging.

There are often times when it is said that *circumstances* or *conditions* are causes or effects. This also is a permissible way of speaking, and here there is no danger of the terms being taken in too narrow a sense, because a static or constant condition is also a condition. Suppose that we investigate the cause of a collision between two cars on a highway. We must study not only the changing conditions—how the cars moved, the behavior of the drivers, and so on—but also the conditions that were constant at the moment of collision. We must find out about the state of the surface of the road. Was it wet or dry? Was the sun shining directly into the face of one of the drivers? Questions such as these can also be important in determining the causes of the crash. To make a full analysis of the causes, we must investigate all relevant conditions, both constant and changing. It may turn out that many different conditions were important contributions to the final result.

When a man dies, a doctor must state the cause of death. He may write "tuberculosis", as if only one thing caused the death. In everyday life, we often demand a single cause for an event—*the* cause of death, *the* cause of the collision. But when we examine the situation more carefully, we see that many answers can be given, depending on the point of view from which the question was raised. A road-building engineer could say: "Well, I have said many times before that this is a poor surface to use for a highway. It gets very slippery when wet. Now we have another accident to prove it!" According to this engineer, the accident was caused by the slippery highway. He is interested in the event from *his* point of view. He singles this out as *the* cause. In one respect, he is right. If his advice had been followed and the road had been given another surface, it would not have been so slippery as it was. Other things being the same, the accident might not have happened. It is difficult to be sure of this in any particular case, but at least there is a good possibility that the engineer is right. When he maintains that "this is the cause", he means: this is an important condition of such a kind that, if it had not been there, the accident might not have occurred.

Other people, when asked about the cause of the accident, may mention other conditions. The traffic police who study the causes of traffic accidents will want to know if either driver violated any rules of the road. Their job is to supervise such activities, and if they find that the rules have been violated, they will refer to that violation as the cause of the crash. A psychologist who interviews one of the drivers may conclude that the driver was in a state of anxiety; so deeply concerned with his worries that he did not give full attention to the approach of the other car at the crossing. The psychologist will say that the man's disturbed state of mind was the cause of the crash. He is picking out the factor in the total situation that most concerns him. For him, this is the interesting, the decisive cause. He, too, may be right, because, if the man had not been in a state of anxiety, the accident might not, or even probably would not, have happened. An automobile construction engineer may find another cause, such as a defect in the structure of one of the cars. A repair-garage man may point out that the brake-lining of one car was worn out. Each person, looking at the total picture from his point of view, will find a certain condition such that he can correctly say: if that condition had not existed, the accident might not have occurred.

None of these men, however, has answered the more general question: what was *the* cause of the accident? They have given only a series of partial answers, pointing out special conditions that contributed to the final result. No single cause can be singled out as *the* cause. Indeed, it is obvious that there is no such thing as *the* cause. There are many relevant components in a complex situation, each contributing to the accident in the sense that, had the component been absent, the crash might not have happened. If a causal relation is to be found between the accident and a previous event, then the previous event must be the *whole* previous situation. When it is said that this earlier situation "caused" the accident, what is meant is that, given the previous situation, in all of its myriad details, and given all the relevant laws, the accident could have been predicted. No one actually knew, of course, or could know, *all* the facts and relevant laws. But *if* someone had known, he could have predicted the collision. "Relevant laws" include not only laws of physics and technology (concerning the friction on the road, the motion of the cars, the operation of the brakes, and so on), but also physiological and psychological laws. Knowledge of all these laws as well as of all the relevant single facts must be presupposed before it can be said that the outcome is predictable.

The result of this analysis can be summed up briefly: *Causal relation means predictability*. This does not mean actual predictability, because no one could have known all the relevant facts and laws. It means predictability in the sense that, *if* the total previous situation had been known, the event could have been predicted. For this reason, when I use the term "predictability" I mean it in a somewhat metaphorical sense. It does not imply the possibility of someone actually predicting the event, but rather a potential predictability. Given all the relevant facts and all the relevant laws of nature, it would have been possible to predict the event before it happened. This prediction is a logical consequence of the facts and laws. In other words, there is a logical relation between the full description of the previous condition, the relevant laws, and the prediction of the event.

The relevant single facts involved in the previous situation can, in principle, be known. (We ignore here the practical difficulty of obtaining all the facts, as well as the limitations imposed in principle by quantum theory on knowing all the facts at the subatomic level.) With respect to knowing the relevant laws, a much larger problem arises. When a causal relation is defined by saying that an event can be logically inferred from a set of facts and laws, what is meant by "laws"? It is tempting to say: This means those laws that can be found in the textbooks of the various sciences involved in the situation; more precisely, all those relevant laws that are known at the time of the event. In formal language, an event Y at the time T is caused by a preceding event X, if and only if Y is deducible from X with the aid of the laws L_T known at the time T.

It is easy to see that this is not a very helpful definition of a causal relation. Consider the following counterexample. There is an historical report of an event B that happened in ancient times, following an event A. People living at the time T_1 could not explain B. Now B can be explained with the help of knowledge of certain laws, L^*, by showing that B follows logically from A and L^*. But at the time T_1 the laws L^* were not known; therefore, the event B could not be explained as the effect of event A. Suppose that at the time T_1 a scientist had asserted, just as a hypothesis, that event B was caused by event A. Looking back, his hypothesis would be said to be true, although the scientist could not prove it. He was unable to prove it because the laws that were known to him, L_{T_1}, did not include the laws L^*, which are essential to the proof. However, if the definition of causal relation suggested in the previous paragraph is accepted, it would be necessary to say that the scientist's assertion is false. It is false, because he is not able to deduce B from A and L_{T_1}. In other words, his assertion must be called false even though it is known today that it is true.

The inadequacy of the proposed definition is also apparent when we reflect on the fact that today's knowledge of the laws of science is also far from complete. Scientists today know more than scientists of any previous period, but they certainly know less than scientists will know (assuming civilization is not destroyed by a holocaust) a hundred years from now. At no time does science possess complete knowledge of all the laws of nature. As shown earlier, however, it is the entire system of laws, rather than just the laws known at a particular time, that must be referred to in order to obtain an adequate definition of causality.

What is meant when it is said that event B is caused by event A? It is that there are certain laws in nature from which event B can be logically deduced when they are combined with the full description of event A. Whether the laws L can be stated or not is irrelevant. Of course, it is relevant if a proof is demanded that the assertion is *true*. But it is not relevant in order to give the *meaning* of the assertion. It is this that makes the analysis of causality such a difficult, precarious task. When a causal relation is mentioned, there is always an implicit reference to unspecified laws of nature. It would be much too exacting, too far out of line with current usage, to demand that every time someone asserted that, "A was the cause of B", he must be able to state all the laws involved. Of course, if he can state all the relevant laws, then he has proved his assertion. But such proof must not be demanded before his statement is accepted as meaningful.

Suppose a wager is made that it will rain four weeks from today. No one knows whether the prediction is right or wrong. It will be four weeks before the question is decided. Nevertheless, the prediction is clearly meaningful. Empiricists are right, of course, when they say that there is no meaning to a statement unless there is, at least in principle, a possibility of finding confirming or disconfirming evidence about the statement. But this does not mean that a statement is meaningful if and only if it is possible to decide *today* about its truth. The prediction of rain is meaningful, even though its truth or falsity cannot be decided now. The assertion that A is the cause of B is also a meaningful assertion, although the speaker may be unable to specify the laws needed to prove the assertion. It means that, *if* all the relevant facts that surround A were known, together with all the relevant laws, the occurrence of B could then be predicted.

This brings up a difficult question. Does this definition of a cause-and-effect relation imply that the effect follows *of necessity* from the cause? The definition does not speak of necessity. It merely says that event B could be predicted if all the relevant facts and laws were known. But perhaps this begs the question. The metaphysician who wishes to introduce necessity into the definition of causality can argue: "It is true that the word 'necessity' is not used. But, laws are spoken of, and laws are statements of necessity. Therefore, necessity comes in after all. It is an indispensable component of any assertion about a causal relation."

In the next chapter we will consider what can be said in reply to this argument.

CHAPTER 20

Does Causality
Imply Necessity?

DO LAWS imply necessity? Empiricists
sometimes formulate their position as follows: a law is merely a universal conditional statement. It is universal because it speaks in a general way. "At any time, at any place, if there is a physical body or
system in a certain state, then another specific state will follow." It is an
if-then statement in general form with respect to time and space. This
approach is sometimes called "conditionalism". A causal law simply
states that, whenever an event of the kind P (P is not a single event but
a class of events) occurs, then an event of the kind Q will follow. In
symbolic form:

(1) $(x) (Px \supset Qx)$

This statement asserts that, at every space-time point x, if P holds,
the condition Q will hold.

Some philosophers object strenuously to this view. They contend
that a law of nature asserts much more than just a universal conditional
statement of the if-then form. To understand their objection, it is necessary to review exactly what is meant by a statement of the conditional

196

form. Instead of the universal statement (1), consider a particular instance of it for the space-time point a.

(2) $Pa \supset Qa$

The meaning of that statement, "If P occurs at a, then Q occurs at a", is given by its truth table. There are four possible combinations of truth values for the two components in the statement:

1. "Pa" is true, "Qa" is true.
2. "Pa" is true, "Qa" is false.
3. "Pa" is false, "Qa" is true.
4. "Pa" is false, "Qa" is false.

The horseshoe mark for implication, "$\supset$", is to be understood in such a way that (2) asserts no more than that the second combination of truth values does not hold. It says nothing about a causal connection between Pa and Qa. If "Pa" is false, the conditional statement holds regardless of whether "Qa" is true or false. And if "Qa" is true, it holds regardless of whether "Pa" is true or false. It fails to hold only when "Pa" is true and "Qa" false.

Obviously, this is not a strong interpretation of a law. When it is said, for example, that iron expands when heated, is nothing more meant than that one event follows the other? It could also be said that, when iron is heated, the earth will rotate. This, too, is a conditional statement, but it would not be called a law, because there is no reason to believe that the earth's rotation has anything to do with the heating of a piece of iron. On the other hand, when a law is stated in conditional form, does it not carry with it a meaning component that asserts some sort of connection between the two events, a connection above and beyond the mere fact that if one occurs the other will follow?

It is true that something more is usually intended when a law is asserted, but exactly what the "more" is, is difficult to analyze. Here we come up against the problem of deciding exactly what constitutes the "cognitive content" of an English statement. Cognitive content is that which is asserted by the statement and is capable of being either true or false. It is often extremely difficult to decide exactly what belongs to the cognitive content of a statement and what belongs to noncognitive meaning components that are there but are irrelevant to the statement's cognitive meaning.

An illustration of this sort of ambiguity is the case of a court witness who says, "Unfortunately, the truck struck Mr. Smith and fractured his left hip." Another witness introduces evidence that clearly shows that the previous witness did not think this "unfortunate" at all. Actually, he was quite pleased to see Mr. Smith injured. Did he or did he not lie when he used the word "unfortunately"? If it is established that the witness did not regret the accident, then clearly his use of the word "unfortunately" was deceptive. From this point of view, it might be said that he lied. But from the point of view of the court, assuming that the statement was made under oath, the question of perjury is a hard one to answer. Perhaps the judge would reason that the use of the word "unfortunately" had no bearing on the real content of the statement. The truck did strike Mr. Smith and did fracture his hip. The witness spoke of this as unfortunate in order to give the impression that he regretted the incident although, in fact, he did not. But this is not relevant to the central assertion of his sentence.

Had the witness said, "Mr. Smith was struck by the truck, and I regret very much that this happened to him", his statement of regret would have been more explicit, and perhaps the question of perjury would be more pertinent. In any case, it is apparent that it is often not easy to decide what belongs to the cognitive content of an assertion and what is merely a factor of noncognitive meaning. The English language has a grammar, but it does not have rules that specify what should and what should not be considered relevant to the truth value of a sentence. If someone says "unfortunately" when he really does not feel regret, is his statement false? There is nothing in an English dictionary or grammar book that will help to answer this question. Linguists can do no more than report on how people in a culture usually take certain statements; they cannot make up rules for deciding the matter in every given case. In the absence of such rules, it is not possible to make a precise analysis of the cognitive content of certain ambiguous statements.

Exactly the same difficulty is involved in trying to decide whether a sentence of the form "$(x)(Px \supset Qx)$" is a complete formulation of a law or whether it leaves out something essential. Ever since philosophers of science began to formulate laws with the help of the symbol "$\supset$", the connective of material implication, voices have been raised against this formulation. To call something a "law of nature", certain philosophers have maintained, is to say much more than that one event follows another. A law implies that the second event *must* follow. There is some sort of *necessary* connection between P and Q. Before this objection can be fully evaluated, we must first find out exactly what these philosophers mean by "necessary", and second, whether this meaning belongs to the cognitive content of the statement of the law.

Many philosophers have tried to explain what they mean by "necessity" when it is applied to laws of nature. One German author, Bernhard Bavink, went so far as to maintain (in his work *Ergebnisse und Probleme der Naturwissenschaften*) that the necessity in laws of nature is a logical necessity. Most philosophers of science would deny this. In my opinion, it is entirely wrong. "Logical necessity" means "logical validity". A statement is logically valid only if it says nothing whatever about the world. It is true merely by virtue of the meanings of the terms occurring in it. But the laws of nature are contingent; that is, for any law, it is quite easy to describe, without self-contradiction, a sequence of processes that would violate it.

Consider the law: "When iron is heated, it expands." Another law says: "When iron is heated, it contracts." There is no logical inconsistency in this second law. From the standpoint of pure logic, it is no more invalid than the first law. The first law is accepted, rather than the second, only because it describes a regularity *observed in nature*. Laws of logic can be discovered by a logician sitting at his desk making marks on paper or just thinking with his eyes closed. No law of nature can be discovered in this manner. Laws of nature can be discovered only by observing the world and describing its regularities. Since a law asserts that a regularity holds for all times, it must be a tentative assertion. It can always be found wrong by a future observation. Laws of logic, however, hold under all conceivable conditions. If there is a necessity in laws of nature, it certainly is not a logical necessity.

What can a philosopher mean, then, when he speaks of necessity in natural law? Perhaps he will say: "I mean that, when P occurs, it cannot possibly be that Q will not follow. It *must* happen. It cannot be otherwise." But such expressions as "must happen" and "cannot be otherwise" are just other ways of saying "necessary", and it is still not clear what he means. He certainly does not wish to reject the conditional statement, "$(x)(Px \supset Qx)$". He agrees that it holds but finds it too weak a formulation. He wants to strengthen it by adding something.

To clarify the issue, assume that there are two physicists, both of whom possess the same factual knowledge and who also agree on the same system of laws. Physicist I makes a list of these laws, expressing all of them in the universal conditional form of $(x)(Px \supset Qx)$. He is satisfied with this formulation and has no desire to add anything. Physicist II makes the same list of laws, expressing them in the same conditional form, but, in each case, he adds, "and this holds with necessity". The two lists will take the following form:

Physicist I

Law 1: $(x)(Px \supset Qx)$
Law 2: $(x)(Rx \supset Sx)$

.
.
.

Physicist II

Law 1: (x) $(Px \supset Qx)$, and this holds with necessity.
Law 2: $(x)(Rx \supset Sx)$, and this holds with necessity.

.
.
.

Is there any difference between these two systems of laws, regarding the cognitive meaning of the two systems? To answer this, it is necessary to try to discover whether there is any test by which the superiority of one system over the other can be established. This, in turn, is the same as asking whether there is any difference in the power of the two systems to predict observable events.

Suppose that the two physicists agree on the present state of the weather. They have access to the same reports from the same weather stations. On the basis of this information, together with their respective systems of laws, they predict the state of the weather tomorrow in Los Angeles. Since they make use of the same facts and the same laws, their predictions will, of course, be the same. Can physicist II, in view of the fact that after each law he has added "and this holds with necessity", make more or better predictions than physicist I? Obviously, he cannot. His additions say nothing whatever about any observable feature of any predicted event.

Physicist I says: "If P, then Q. Today there is P; therefore, tomorrow there will be Q." Physicist II says: "If P, then Q, and this holds with necessity. Today there is P; therefore, tomorrow there will be Q, say, a thunderstorm. But, not only will there be a thunderstorm in Los Angeles tomorrow, there *must* be a thunderstorm." Tomorrow comes. If there is a thunderstorm, both physicists are pleased with their success. If there is no thunderstorm, they will both say: "Let's see if we can find the source of our error. Perhaps the reports were incomplete or faulty. Perhaps one of our laws is wrong." But is there any basis on which physicist II can make a prediction that cannot also be made by physicist I? Obviously not. The additions made by the second physicist to his list of laws are completely without influence on the ability to make predictions. He believes that his laws are *stronger,* that they say more, than the laws of his rival. But they are stronger only in their ability to arouse an emotional feeling of necessity in the mind of the second physicist. They are certainly not stronger in their cognitive meaning, because the cognitive meaning of a law lies in its potentialities for prediction.

Not only is it true that no more can be predicted by the laws of physicist II in any actual test, but no more can be predicted *in principle.* Even if we assume hypothetical weather conditions—strange conditions that never occur on the earth but can be imagined—the two physicists would still make identical predictions on the basis of identical facts and their respective lists of laws. For this reason, the modern empiricist takes the position that the second physicist has added nothing significant to his laws.

This is essentially the position taken by David Hume in the eighteenth century. In his famous critique of causality, he argued that there is no basis for assuming that an intrinsic "necessity" is involved in any observed cause-and-effect sequence. You observe event *A,* then you observe event *B.* What you have observed is no more than a temporal succession of events, one after the other. No "necessity" has been observed. If you don't observe it, Hume said in effect, don't assert it. It adds nothing of value to the description of your observations. Hume's analysis of causality may not have been entirely clear or correct in all details, but, in my opinion, it was essentially correct. Moreover, it had the great merit of focusing the attention of later philosophers on the inadequacy with which causality had previously been analyzed.

Since the time of Hume, the most important analyses of causality, by Mach, Poincaré, Russell, Schlick, and others, have given stronger and stronger support to Hume's conditionalist view. A statement about a causal relation is a conditional statement. It describes an observed regularity of nature, nothing more.

Let us turn now to another aspect of causality, an important respect in which a causal relation differs from other relations. In most cases, in order to determine whether a relation *R* holds between an event or object *A* and an event or object *B,* we simply study *A* and *B* carefully to see if the relation *R* obtains. Is building *A* taller than building *B?*

We inspect the two buildings and reach a conclusion. Is wallpaper *C* a darker shade of blue than wallpaper *D?* It is not necessary to examine other samples of wallpaper to answer this question. We study *C* and *D*, under normal lighting, and reach a decision on the basis of our understanding of what is meant by "darker shade of blue". Is *E* a brother of *F?* Perhaps they do not know whether they are brothers. In this case we must study their antecedent history. We go back into their past and try to determine whether they had the same parents. The important point is that there is no need to study other cases. We examine only the case at hand to determine whether a certain relation holds. Sometimes this is easy to determine, sometimes extremely difficult, but it is not necessary to examine other cases to decide whether the relation holds for the case in question.

With respect to a causal relation, this is not so. To determine whether a certain causal relation holds between *A* and *B*, it is not sufficient merely to define a relation and then study the pair of events. That is, theoretically, it is not sufficient. In actual practice, because we possess a great deal of knowledge about other events, it is not always necessary to examine other events before saying that a causal relation holds between *A* and *B*. The relevant laws may be so obvious, so familiar, that they are tacitly assumed. It is forgotten that these laws are accepted only because of many previous observations of cases in which the causal relation held.

Suppose I see a stone moving toward a window, striking the pane, and then the glass splintering into a thousand pieces. Was it the impact of the stone that caused the destruction of the pane? I say that it was. You ask: how do you know this? I reply: it was obvious. I saw the stone hit the window. What else could have caused the glass to break? But note that my very phrase, "what else", raises a question of knowledge concerning other events in nature similar to the event in question. From early childhood we have observed hundreds of cases in which glass was shattered by a strong impact of some sort. We are so accustomed to this sequence of events that when we see a stone moving toward a window we anticipate the breaking of the glass even before it happens. The stone hits the pane. The pane shatters. We take for granted that the impact of the stone caused the shattering.

But think how easy it is to be deceived by appearances. You watch a TV western movie and see the villain point his pistol at another man and pull the trigger. The sound of a shot is heard, and the other man falls down dead. Why did he fall? Because he was struck by a bullet. But there was no bullet. Even the sound of the shot may have been dubbed in later onto the film's sound track. The causal sequence you thought you observed was entirely illusory. It was not there at all.

In the case of the stone and window, perhaps the stone struck a hard, invisible plastic surface in front of the window. The surface did not break. However, just as the stone hit this surface, someone inside the house, to deceive you, shattered the window by some other means. It is possible, then, to be deceived, to believe that a causal relation holds when, in fact, it does not. In this case, however, such deceptions are ruled out as improbable. Experience of similar events in the past makes it likely that this is another instance of glass being shattered by a moving object. If there is a suspicion of deception, a more thorough study of the case is made.

The essential point here is this: whether we observe the case superficially and conclude that the stone did, in fact, break the glass, or whether we suspect deception and study it in more detail, we are always doing more than studying just the one case. We are bringing to bear upon it many hundreds of other cases of a similar nature that have been experienced in the past. It is never possible to assert a causal relation on the basis of observing one case alone. As children we see things happen in temporal sequences. Gradually, over the years, we form impressions of certain regularities that occur in our experience. A drinking glass is dropped. It breaks. A baseball strikes the window of a car. The window cracks. In addition, there are hundreds of similar experiences in which fragile material similar to glass, a porcelain saucer, for example, is shattered by a blow. Without such experiences, the observation of the stone and windowpane would not be interpreted as a causal relation.

Suppose that, at some future date, all window glass is made so that it can be shattered only by an extremely high frequency sound. If this knowledge provided the background of our experience, and we saw a windowpane shatter when a stone struck it, we would exclaim: "What a strange coincidence! At the very instant that the stone struck the glass, someone inside the building produced a high frequency sound that shattered the glass!" It is apparent, then, that a peculiar feature of the causal relation, distinguishing it from other relations, is that it cannot be established by the inspection of only one concrete case. It can be established only on the basis of a general law, which, in turn, is based on many observations of nature.

When someone asserts that *A* caused *B,* he is really saying that this is a particular instance of a general law that is universal with respect to space and time. It has been observed to hold for similar pairs of events, at other times and places, so it is assumed to hold for any time or place. This is an extremely strong statement, a bold leap from a series of particular instances to the universal conditional: for every *x*, if *Px* then *Qx*. If *Pa* is observed, then, together with the law, *Qa* logically follows. The law could not be asserted had there not been many previous observations; this is the way in which the causal relation is fundamentally different from other relations. In the case of the relation, "object *x* is inside box *y*", one examination of a particular box *b* is sufficient to determine whether a particular object *a* is inside. But, to determine whether the cause-effect relation holds in a particular instance, it is not sufficient to examine that one instance. A relevant law must first be established, and this requires repeated observations of similar instances.

From my point of view, it is more fruitful to replace the entire discussion of the meaning of causality by an investigation of the various kinds of laws that occur in science. When these laws are studied, it is a study of the kinds of causal connections that have been observed. The logical analysis of laws is certainly a clearer, more precise problem than the problem of what causality means.

To understand causality from this modern point of view, it is instructive to consider the concept's historical origin. I have made no studies of my own in this direction, but I have read with interest what Hans Kelsen has written about it.[1] Kelsen is now in this country, but, at one time, he was a professor of constitutional and international law at the University of Vienna. When the revolution came in 1918 and the Austrian Republic was founded the following year, he was one of the main authors of the Republic's new constitution. In analyzing philosophical problems connected with law, he apparently became interested in the historical origins of the concept of causality.

It is often said that there is a tendency for human beings to project their own feelings into nature, to suppose that natural phenomena like rain and wind and lightning are animated and act with purposes similar to those of a human. Is this, perhaps, the origin of the belief that there are "forces" and "causes" in nature? Kelsen became convinced that this

---

[1] Kelsen's views are expressed in his paper, "Causality and Retribution," *Philosophy of Science*, 8 (1941), and developed in greater detail in his book, *Society and Nature* (Chicago, Ill.: University of Chicago Press, 1943).

analysis of the origin of the concept of causality, plausible though it seems, is too individualistic. In his studies of the concept's first appearance, in ancient Greece, he found that the social order, not the individual, served as the model. This is suggested by the fact that, from the beginning, and even today, the regularities of nature are called "laws of nature", as if they were similar to laws in the political sense.

Kelsen explained it this way. When the Greeks began their systematic observations of nature and noticed various causal regularities, they felt that a certain necessity was behind the phenomena. They looked on this as a moral necessity analogous to the moral necessity in relations between persons. Just as an evil deed demands punishment and a good deed demands reward, so a certain event A in nature demands a consequent B to restore a harmonious state of things, to restore justice. If it gets colder and colder in the autumn, reaching an extreme of cold in winter, then the weather, so to speak, gets out of balance. To restore the balance, the rightness of things, the weather must now grow warmer and warmer. Unfortunately, it goes to the other extreme and gets too hot, so the cycle must be repeated. When nature moves too far away from a balanced, harmonious state of affairs, analogous to the harmonious society, the balance must be restored by an opposite trend. This concept of a natural order or harmony reflected the Greek love of social order and harmony, their love of moderation in all things, their avoidance of extremes.

Consider the principle that cause and effect must in some way be equal. The principle is embodied in many physical laws, such as Newton's law that action is accompanied by an equal reaction. It has been stressed by many philosophers. Kelsen believes that this was originally an expression of the social belief that a punishment must equal the crime. The more atrocious the crime, the more severe the punishment. The greater the good deed, the greater the reward. Such a feeling, grounded in a social structure, was projected upon nature and became a basic principle of natural philosophy. "Causa aequat effectum", the medieval philosophers expressed it. Among metaphysical philosophers today, it still plays an important role.

I remember a discussion I once had with a man who said that the Darwinian theory of evolution could be rejected completely on metaphysical grounds. There was no way, he maintained, by which lowly organisms, possessing a very primitive quality of organization, could develop into higher organisms, with higher organization. Such a development would violate the principle of the equality of cause and effect. Only Divine interference could account for the change. Belief in the *causa aequat effectum* principle was so strong for this man that he rejected a scientific theory because he supposed that it violated the principle. He did not attack the theory of evolution by evaluating its evidence. He simply rejected it on metaphysical grounds. Organization cannot come from nonorganization, because causes must equal effects; a higher Being must be invoked to explain evolutionary development.

Kelsen supports his viewpoint with some interesting quotations from Greek philosophers. Heraclitus, for example, speaks of the sun as moving through the sky in obedience to "measures", by which the philosopher means the prescribed limits of its path. "The sun will not overstep his measures", Heraclitus writes, "but if he does, the Erinyes, the handmaidens of Dike, will find him out." The Erinyes were the three demons of revenge, and Dike was the goddess of human justice. The regularity of the sun's path, then, is explained in terms of the sun's obedience to a moral law decreed by the gods. If the sun disobeys and steps out of line, retribution will catch up with him.

On the other hand, there were some Greek philosophers who strongly opposed this view. Democritus, for example, regarded the regularities of nature as completely impersonal, not connected in any way with divine commands. He probably thought of these laws as possessing an intrinsic, metaphysical necessity; nevertheless, this step from the personal necessity of divine commands to an impersonal, objective necessity was a great step forward. Today, science has eliminated the concept of metaphysical necessity from natural law. But, in Democritus' time, his view was an important advance over the view of Heraclitus.

In Philipp Frank's book on causality, *Das Kausalgesetz und seine Grenzen* (published in Vienna in 1932 and not translated into English), he points out that it is often instructive to read the prefaces of scientific textbooks. In the body of such a book, the author may be entirely scientific, careful to avoid all metaphysics. But prefaces are more personal. If the author has a hankering for the older, metaphysical way of looking at things, he may feel that his preface is the proper place to tell his readers what science is *really* all about. Here you may discover what sort of philosophic notions the author had in the back of his head when he wrote his textbook. Frank quotes from the preface of a contemporary physics text: "Nature never violates the laws." This seems innocent enough, but when it is carefully analyzed, it is seen to be a most curious remark. What is curious is not the belief in causality, but the way in which it is expressed. He does not say that sometimes there are miracles, exceptions to causal law. In fact, he explicitly denies this. But he denies it by saying that nature never *violates* the laws. His words imply that nature has some sort of choice. Certain laws are given to nature. Nature could, from time to time, violate one of them; but, like a good, lawful citizen, she never does. If she did, presumably the Erinyes would arrive on the scene and set her back on the right path. You see, there still lingers here the notion of laws as commands. The author would, of course, be insulted if you attributed to him the old metaphysical view that laws are given to nature in such a way that nature can obey or disobey. But, from the way he chose his words, the old point of view must still be lingering in his mind.

Suppose that on visiting a city for the first time you use a street map to help you find your way about. Suddenly you discover a clear discrepancy between the map and the city's streets. You do not say, "The streets are disobeying the law of the map." Instead you say, "The map is wrong." This is precisely the situation of the scientist with respect to what are called the laws of nature. The laws are a map of nature drawn by the physicists. If a discrepancy is discovered, the question is never whether *nature* disobeyed; the only question is whether the *physicists* made an error.

Perhaps it would be less confusing if the word "law" were not used at all in physics. It continues to be used, because there is no generally accepted word for the kind of universal statement that a scientist uses as the basis for prediction and explanation. In any case, it should be kept clearly in mind that, when a scientist speaks of a law, he is simply referring to a description of an observed regularity. It may be accurate, it may be faulty. If it is not accurate, the scientist, not nature, is to blame.

CHAPTER 21

The Logic of
Causal Modalities

BEFORE GOING DEEPER into the nature of scientific laws, I should like to clarify some previous brief remarks on Hume. I believe that Hume was right in saying that there is no intrinsic necessity in a causal relation. However, I do not deny the possibility of introducing a necessity concept, provided it is not a metaphysical concept but is a concept within the logic of modalities. Modal logic is a logic that supplements the logic of truth values by introducing such categories as necessity, possibility, and impossibility. Great care must be taken to distinguish among logical modalities (logically necessary, logically possible, and so on) and causal modalities (causally necessary, causally possible, and so on), as well as many other kinds of modalities. Only the logical modalities have been studied extensively. The best known work in this field is the system of strict implication developed by C. I. Lewis. I myself once published a paper on this topic. But in reference to the causal relation, it is not logical modality but causal modality with which we must be concerned.

In my opinion, a logic of causal modalities is possible. So far, very little work has been done in this field. The first attempt to develop a system of this type seems to have been by Arthur W. Burks.[1] He proposes an axiom system, but an extremely weak one. Actually, he does not specify under what conditions a universal statement would be regarded as causally necessary. Others have attacked essentially the same problem but in a different terminology. For instance, Hans Reichenbach has done so in his little book, *Nomological Statements and Admissible Operations*.[2] A great many articles have dealt with the problem of "counterfactual conditionals", a problem closely connected with this one.

A counterfactual conditional is an assertion that, if a certain event had not taken place, then a certain other event would have followed. Obviously, the meaning of this assertion cannot be conveyed in a symbolic language by using the truth-functional conditional (the symbol "⊃") in the sense in which it is ordinarily used. An attempt to analyze the precise meaning of counterfactual-conditional statements raises a variety of difficult problems. Roderick M. Chisholm (1946) and Nelson Goodman (1947) were among the first to write about this.[3] Since then, many authors have followed with other papers.

Exactly what is the connection between the problem of counterfactual conditionals and the problem of formulating a modal logic that will include the concept of causal necessity? The connection lies in the fact that a distinction must be made between two kinds of universal statements. On the one hand, there are what may be called genuine laws, such as the laws of physics, which describe regularities universal in space and time. On the other hand, there are universal statements, which are not genuine laws. Various terms have been proposed for them; sometimes they have been called "accidental" universals. An example is, "All the coins in my pocket on January 1, 1958, were silver". The essential difference between the two kinds of universal

[1] See Burks's paper, "The Logic of Causal Propositions," *Mind*, 60 (1951), 363–382.
[2] Hans Reichenbach, *Nomological Statements and Admissible Operations* (Amsterdam: North-Holland Publishing Co., 1954); reviewed by Carl G. Hempel, *Journal of Symbolic Logic*, 20 (1956), 50–54.
[3] On counterfactual conditionals see Chisholm's paper, "The Contrary-to-Fact Conditional," *Mind*, 55 (1946), 289–307, reprinted in Herbert Feigl and Wilfrid Sellars, eds., *Readings in Philosophical Analysis* (New York: Appleton-Century-Crofts, 1953), and Nelson Goodman's "The Problem of Counterfactual Conditionals," *Journal of Philosophy*, 44 (1947), 113–128, reprinted in his *Fact, Fiction, and Forecast* (Cambridge: Harvard University Press, 1955). Ernest Nagel discusses the topic in his *The Structure of Science* (New York: Harcourt, Brace and World, 1961), pp. 68–73, and cites more recent references.

statements can best be understood by considering counterfactual statements relating to them.

Consider first a genuine law, the law of gravitation. It permits me to assert that, if I drop a stone, it will fall toward the earth with a certain acceleration. I can make a similar statement in counterfactual form by saying: "Yesterday I held a stone in my hand. But if I had not held it, that is, if I had withdrawn my hand, it would have fallen to the earth." This statement does not describe what actually happened, but what *would* have happened, if I had not held the stone. I make this assertion on the basis of the law of gravitation. The law may not be explicitly invoked, but it is tacitly assumed. By stating the law, I give my reason for believing in the counterfactual statement. Clearly, I do not believe it because I saw it happen. It did not happen. But it is reasonable to assert the counterfactual because it is based on a genuine law of physics. The law is considered sufficient justification for the counterfactual.

Can the same be done with the second type of universal statement, the accidental universal? It is apparent at once that it would be absurd. Suppose I say, "If this penny here had been in my pocket on the first of January, 1958, it would have been made of silver." Clearly, the substance of this penny is not dependent on whether or not I had it in my pocket on certain dates. The universal statement, "All coins in my pocket on January 1, 1958, were silver", is not an adequate basis for asserting a counterfactual. It is evident, then, that some universal statements furnish a reasonable basis for a counterfactual, while other universal statements do not. We may be convinced that a universal statement of the accidental type is true; nevertheless, we would not regard it as law. It is essential to keep this distinction in mind when the meaning of counterfactuals is analyzed. It is also involved in the problem of nonlogical, or causal, modalities.

The guiding idea in my approach to the problem is as follows. Let us assume someone proposes a statement as a new law of physics. It is not known whether the statement is true or false, because the observations made so far are insufficient; but it is universal, because it says that, if a certain event occurs at any time or place, a certain other event will follow. By inspecting the *form* of the statement, it can be decided whether the statement would be called a genuine law *if* it were true. The question of whether the law is or is not true is irrelevant; the point is only whether it has the *form* of a genuine law. For instance, someone proposes a law of gravitation that says the force of gravity diminishes with the third power of the distance. This is obviously false; that is, in this universe, it does not hold. But it is easy to conceive of a universe in which it would hold. Therefore, instead of classifying statements into nomological statements or genuine laws (which implies that they are true) and non-nomological statements, I prefer to divide statements, regardless of their truth, into these two classes: (1) statements that have a lawlike form (sometimes called "nomic form") and (2) statements that do not. Each class includes true and false statements. The statement, "Gravity decreases with the third power of the distance", is of the first kind. It is lawlike even though not true and, therefore, not a law. The statement, "On January 1, 1958, all the men in Los Angeles wore purple neckties", is of the second kind. Even if it happened to be true, it would still not express a law but only an accidental state of affairs at one particular time.

It is my conviction that the distinction between these two kinds of statements can be precisely defined. This has not yet been done, but, if it were done, I have a hunch—I will not put it more strongly—that it would be a purely semantic distinction. What I mean is that, if someone presents me with a universal statement *S* and if I have made the distinction between the two types sufficiently clear to myself, I would not have to perform any experiments to decide which kind the statement is. I would merely ask myself: If the world were such that *S* is true, would I regard it as a law? To put the question more precisely: Would I regard it as a *basic* law. Later I will explain my reason for making this distinction. Now I wish only to make clear what I mean by "having the form of a possible basic law", or, more briefly, "having *nomic form*".

The first condition for a statement having nomic form was made clear by James Clerk Maxwell, who, a century ago, developed the classic electromagnetic theory. He pointed out that the basic laws of physics do not speak of any particular position in space or point in time. They are entirely general with respect to space and time; they hold everywhere, at all times. This is characteristic only of *basic* laws. Obviously there are many important technical and practical laws that are not of this kind. They stand in a position that is between basic laws and accidentals, but they are not entirely accidental. For example: "All the bears in the North Polar region are white." This is not a basic law, because the facts could be quite otherwise. On the other hand, neither is it quite accidental; certainly, it is not so accidental as the fact that all the coins in my pocket were silver on a certain date. The statement about the polar bears is dependent on a variety of basic laws that determine the climate near the North Pole, the evolution of bears, and other factors. The color of the bears is not accidental. On the other hand, the climate may change during the next million years. Other species of bears, with different-colored fur, may evolve near the Pole or move there. The statement about the bears cannot, therefore, be called a basic law.

Sometimes a law is thought to be basic but later proves to be limited to a time or place or to certain conditions. Nineteenth-century economists spoke of laws of supply and demand as though they were general economic laws. Then the Marxists came along with their criticisms, pointing out that these laws were true for only a certain type of market economy but were in no sense laws of nature. In many fields—biology, sociology, anthropology, economics—there are laws that seem at first to hold generally, but only because the author did not look beyond the limits of his country, or his continent, or his period of history. Laws thought to express a universal moral behavior or universal forms of religious worship turned out to be limited laws when it was discovered that other cultures behaved differently. Today, it is suspected that there may be life on other planets. If so, many laws of biology, which are universal with respect to living things on earth, may not apply to life elsewhere in the galaxy. Apparently, then, there are many laws that are not accidental, but that hold only in certain limited regions of space-time and not universally. It is necessary to distinguish between those laws and the universal laws. The laws called the laws of physics are believed to hold everywhere. Maxwell, when he formulated his equations for electromagnetism, was convinced that they obtained not only in his laboratory, but also in any laboratory, and not only on the earth, but also in space and on the moon and on Mars. He believed he was formulating laws that were universal throughout the universe. Although his laws have been somewhat modified by quantum mechanics, they have only been modified. In essential respects, they are still regarded as universal, and, whenever a modern physicist states a basic law, he intends it to be universal. Such basic laws must be distinguished from spatio-temporally restricted laws and from derivative laws that hold only for certain kinds of physical systems, certain substances, and so forth.

The problem of defining precisely what is meant by nomic form, that is, the form of a possible basic law, has not yet been settled. Certainly Maxwell's condition that the law apply to all times and places must be part of the definition. There should be other conditions. Several have been proposed, but philosophers of science are not yet agreed on exactly what these additional conditions should be. Putting aside this unsolved problem, let us assume that there is an exact definition of nomic form. I shall now indicate how, in my view, that nomic form can provide the basis for defining some other important concepts.

First, I define a *basic law* of nature as a statement that has nomic form and is also true. The reader may feel uneasy about this definition. Some of my friends contended that an empiricist should never speak about a law being true; a law refers to infinitely many instances, throughout all space and time, and no human being is ever in a position to know with certainty whether it holds universally or not. I agree. But a clear distinction must be made between certainty and truth. There is never, of course, any certainty. Indeed, there is less certainty with respect to a basic law than to a singular fact. I am more certain that this particular pencil has just dropped from my hand to the desk than I am about the universality of laws of gravitation. That does not, however, prevent one from speaking meaningfully of a law being true or not true. There is no reason why the concept of truth cannot be used in defining what is *meant* by a basic law.

My friends argued that they would prefer to say, instead of "true", "confirmed to a high degree". Reichenbach, in his book *Nomological Statements and Admissible Operations,* cited earlier, comes to the same conclusion, although in different terminology. By "true", he means "well established" or "highly confirmed on the basis of available evidence at some time in the past, present, or future". But this is not, I suspect, what scientists *mean* when they speak of a basic law of nature. By "basic law", they mean something that holds in nature regardless of whether any human being is aware of it. I am convinced that this is what most writers of the past as well as most scientists today mean when they speak of a law of nature. The problem of defining "basic law" has nothing to do with the degree to which a law has been confirmed; such confirmation can never, of course, be complete enough to provide certainty. The problem is only concerned with the meaning that is intended when the concept is used in the discourse of scientists.

Many empiricists become uneasy when they approach this question. They feel that an empiricist should never use a terribly dangerous word like "true". Otto Neurath, for instance, said that it would be a sin against empiricism to speak of laws as true. American pragmatists, including William James and John Dewey, held similar views. In my opinion, this judgment is explained by a failure to distinguish clearly between two different concepts: (1) the degree to which a law is established at a certain time and (2) the semantic concept of the truth of a law. Once this distinction is made and it is realized that, in semantics, a precise definition of truth can be provided, there is no longer any reason for hesitating to use the word "truth" in defining a "basiç law of nature".

I would propose the following definition: a statement is *causally true,* or C-true, if it is a logical consequence of the class of all basic laws. Basic laws are defined as statements that have nomic form and are true. Those C-true statements that have universal form are laws in the wider sense, either basic laws or derivative laws. The derivative laws include those restricted in space and time, such as the laws of meteorology on the earth.

Consider the following two statements. "In the town of Brookfield, during March 1950, on every day when the temperature stood below the freezing point from midnight to five A.M., at five A.M. the town pond was covered with ice." This is a derivative law. Compare it with the second statement, which runs like the first, except at the end: ". . . then, in the afternoon, a football game took place in the stadium". This statement is also true. There was a football game every Saturday, and the specified temperature condition happened to be fulfilled only twice in March 1950, both times on a Saturday morning. Thus, the second statement, although true and possessing the same logical form as the first, is not a law. It is merely an accidental universal. This example shows that among *restricted* statements of universal form, although assumed true, the distinction between laws (in this case derivative) and accidental universals cannot be made on the sole basis of a semantic analysis of the statements. In my opinion, this distinction can be made only indirectly, with the help of the concept of basic law. A derivative law is a logical consequence of the class of basic laws; the accidental statement is not. However, the distinction between the forms of basic laws and accidental universals can be made, I think, by a purely semantic analysis, without the use of factual knowledge.

In my book *Meaning and Necessity*[4] I defend the view that logical modalities are best interpreted as properties of propositions, analogous

---

[4] Rudolf Carnap, *Meaning and Necessity: A Study in Semantics and Modal Logic* (Chicago: University of Chicago Press, 1947); enlarged edition, hardcover (1956), paperback (1960).

to certain semantic properties of statements that express those propositions. Suppose that a statement S_1 in a language L expresses the proposition p_1; then p_1 is a logically necessary proposition if and only if S_1 is L-true in language L (I use the term "L-true" for "logically true"). The following two statements are therefore equivalent:

(1) S_1 is L-true (in L).
(2) p_1 is logically necessary.

In other words, to say that a proposition is logically necessary is the same as saying that any statement expressing the proposition is L-true. The semantic L-concepts (L-truth, L-falsity, L-implication, L-equivalence) can be defined for languages that are sufficiently strong to contain all mathematics and physics, so the problem of the interpretation of *logical* necessity has been solved. The best approach to other modalities, in particular, to causal modalities, is, in my view, by way of analogy with this one.

As an example of what I mean, consider the difference between statements (1) and (2) above. "S_1" is the name of a sentence, therefore, (1) is a statement in the metalanguage. On the other hand, (2) is an object language statement, although not in an extensional object language. It is an object language with connectives that are not truth-functions. To put sentence (2) in symbolic form, write:

(3) $N(p_1)$

This means "p_1 is a logically necessary proposition."

In an analogous fashion, I would first define "nomic form", then "basic law", and finally "C-true" (causally true). These are all semantic concepts. Thus if we have the statement:

(4) S_1 is C-true,

I would say that the proposition expressed by S_1 is necessary in a causal sense. This can be written as:

(5) p_1 is causally necessary.

Or, in symbolic form:

(6) $N_C(p_1)$

As I define the terms, the class of causally necessary propositions is comprehensive. It contains the logically necessary propositions. In my view, this is more convenient than other ways of defining the same terms, but it is, of course, merely a matter of convenience. The subject of causal modalities has not been much investigated. It is a vast, complicated topic, and we shall not go into any further technicalities here.

CHAPTER 22

Determinism and
Free Will

.

"CAUSALITY" and "causal structure of the world" are terms I prefer to use in an extremely wide sense. Causal laws are those laws by which events can be predicted and explained. The totality of all these laws describes the causal structure of the world.

Of course, everyday speech does not speak of A causing B unless B is later in time than A and unless there is a direct line of causal events from A to B. If a human footprint is seen on the sand, it can be inferred that someone walked across the sand. It would not be said that the footprint *caused* someone to walk across the sand, even though the walking can be inferred from the footprint on the basis of causal laws. Similarly, when A and B are the end results of long causal chains that trace back to a common cause, it is not said that A *caused* B. If it is daytime, the arrival of night can be predicted because day and night have a common cause, but it is not said that one causes the other. After looking at a timetable, it can be predicted that a train will arrive at a certain time; the entry in the table is not thought to cause the train's arrival. Here again, the two events trace back to a common cause. A decision by the management of the railroad company started two separate chains of causally related events that culminated in *A* and *B*. When we read the timetable we make a causal inference that traces back along one chain and forward along the other, but this is such an indirect process that we do not say *B* is caused by *A*. Nevertheless, the process is a causal inference. There is no reason why the term "causal law" cannot be used in a comprehensive way that applies to all the laws by which certain events are predicted and explained on the basis of other events, regardless of whether inferences go forward or back in time.

In the context of this point of view, what can be said about the meaning of the term "determinism"? In my opinion, determinism is a special thesis about the causal structure of the world. It is a thesis that maintains that this causal structure is so strong that, given a complete description of the entire state of the world at one instant in time, then with the help of the laws, any event in the past or future can be calculated. This was the mechanistic view held by Newton and analyzed in detail by Laplace. It includes, of course, within the description of an instantaneous state of the world, not only a description of the position of every particle in the world, but also of its velocity. *If* the causal structure of the world is strong enough to permit this thesis—and I have stated the thesis as Laplace stated it—it can be said that this world has not only a causal structure, but, more specifically, a *deterministic structure*.

In present-day physics, quantum mechanics has a causal structure that most physicists and philosophers of science would describe as not deterministic. It is, so to speak, weaker than the structure of classical physics because it contains basic laws that are essentially probabilistic; they cannot be given a deterministic form like: "If certain magnitudes have certain values, then certain other magnitudes have exactly specified other values." A statistical or probabilistic law says that if certain magnitudes have certain values, there is a specific probability distribution of the values of other magnitudes. If some basic laws of the world are probabilistic, the thesis of determinism does not hold. Today, it is true that most physicists do not accept determinism in the strict sense in which the term has been used here. Only a small minority believe that physics may some day come back to it. Einstein himself never abandoned this belief. He was convinced throughout his life that the present rejection of determinism in physics is only a temporary phase. At present, it is not known whether Einstein was right or wrong.

The problem of determinism is, of course, closely connected in the history of philosophy with the problem of free will. Can a man choose between different possible actions, or is his feeling that he has freedom of choice a delusion? No detailed discussion of this question will be given here, because, in my opinion, it is not affected by any of the fundamental concepts or theories of science. I do not share Reichenbach's opinion that, if physics had retained the classical position of strict determinism, we could not meaningfully speak of making a choice, uttering a preference, making a rational decision, being held responsible for our acts, and so on. I believe that all those things are entirely meaningful, even in a world that is deterministic in the strong sense.[1]

The position I reject—the position held by Reichenbach and others—can be summarized as follows. If Laplace is right—that is, if the entire past and future of the world is determined by any given temporal cross section of the world—then "choice" has no meaning. Free will is an illusion. We think we have a choice, that we make up our minds; actually, every event is predetermined by what happened before, even before we were born. To restore meaning to "choice", therefore, it is necessary to look toward the indeterminacy of the new physics.

I object to this reasoning, because I think it involves a confusion between determination in the theoretical sense, in which an event is determined by a previous event according to laws (which means no more than predictability on the basis of observed regularities), and compulsion. Forget for the moment that, in present-day physics, determinism in the strongest sense does not hold. Think only of the nineteenth-century view. The commonly accepted view of physics was that stated by Laplace. Given an instantaneous state of the universe, a man who possessed a complete description of that state, together with all the laws (of course there is no such man, but his existence is assumed), then he could calculate any event of the past or future. Even if this strong view of determinism holds, it does not follow that the laws *compel* anyone to act as he does. Predictability and compulsion are two entirely different things.

---

[1] A detailed discussion of this question, from a standpoint with which I agree, can be found in "The Freedom of the Will," Chapter 6 of *Knowledge and Society,* a volume written by several Berkeley professors, G. P. Adams and others, "the University of California Associates" (New York: Appleton-Century Co., 1938). The authors of the individual chapters are not identified; but I understand that the late Paul Marhenke was the principal author—perhaps the sole author—of Chapter 6. Since the main points of the chapter are in good agreement with the views of Moritz Schlick, who was a visiting professor at Berkeley several years before the publication of this book, I believe that the chapter shows the effect of his influence.

To explain this, consider a prisoner in a cell. He would like to escape, but he is surrounded by thick walls and the door is locked. This is real compulsion. It can be called negative compulsion because it restrains him from doing something he wants to do. There also is positive compulsion. I am stronger than you and you have a pistol in your hand. You may not want to use it, but if I grab your hand, point the pistol at someone, and forcibly press your finger until it pulls the trigger, then I have compelled you to shoot, to do something you did not wish to do. The law will recognize that I, not you, am responsible for the shooting. This is positive compulsion in a narrow physical sense. In a wider sense, one person can compel another by all sorts of nonphysical means, such as by threatening terrible consequences.

Now compare compulsion, in these various forms, with determination in the sense of regularities occurring in nature. It is known that human beings possess certain character traits which give a regularity to their behavior. I have a friend who is extremely fond of certain musical compositions by Bach that are seldom performed. I learn that a group of excellent musicians are giving a private performance of Bach, at the home of a friend, and that some of these compositions are on the program. I am invited and told I may bring someone. I call my friend, but before I do this, I am almost certain that he will want to go. Now on what basis do I make this prediction? I make it, of course, because I know his character traits and certain laws of psychology. Suppose that he actually comes with me, as I had expected. Was he compelled to go? No, he went of his own free will. He is never freer, in fact, than when given a choice of this sort.

Someone asks him: "Were you compelled to go to this concert? Did anyone exert any sort of moral pressure on you, such as telling you that the host or the musicians would be offended if you did not come?"

"Nothing of the kind", he replies. "No one exerted the slightest pressure. I am very fond of Bach. I wanted very much to come. That was the reason I came."

The free choice of this man is surely compatible with the view of Laplace. Even if total information about the universe, prior to his decision, made it possible to predict that he would attend the concert, it still could not be said that he went under compulsion. It is compulsion only if he is forced by outside agents to do something contrary to his desire. But if the act springs from his own character in accordance with the laws of psychology, then we say that he acted freely. Of course his character is molded by his education, by all the experiences he has had since he was born, but that does not prevent us from speaking of free choices if they spring from his character. Perhaps this man who liked Bach also liked to take walks in the evening. On this particular evening he wished to hear Bach even more than to take his walk. He acted according to his own system of preferences. He made a free choice. This is the negative side of the question, a rejection of the notion that classical determinism would make it impossible to speak meaningfully of free human choices.

The positive side of the question is equally important. Unless there is causal regularity, which need not be deterministic in the strong sense, but may be of a weaker sort, unless there is some causal regularity, it is not possible to make a free choice at all. A choice involves a deliberate preference for one course of action over another. How could a choice possibly be made if the consequences of alternative courses of action could not be foreseen? Even the simplest choices depend on foreseeing possible consequences. A drink of water is taken because it is known that, according to some laws of physiology, it will assuage thirst. The consequences are, of course, known only with varying degrees of probability. Even if the universe is deterministic in the classical sense, this is still true. Sufficient information to be able to predict with certainty is never available. The imaginary man in Laplace's formulation can make perfect predictions, but no such man exists. The practical situation is that knowledge of the future is probabilistic, regardless of whether determinism does or does not hold in the strong sense. But in order to make any sort of free choice, it must be possible to weigh the probable results of alternative courses of action; this could not be done unless there was sufficient regularity in the causal structure of the world. Without such regularities, there would be no moral responsibility or legal responsibility. A person who is unable to foresee the consequences of an act certainly could not be held responsible for that act. A parent, a teacher, a judge regards a child as responsible only in those situations in which the child can foresee the consequences of his acts. Without causality in the world, there would be no point in educating people, in making any sort of moral or political appeal. Such activities make sense only if a certain amount of causal regularity in the world is presupposed.

These views may be summarized in this fashion. The world has a causal structure. It is not known whether this structure is deterministic in the classical sense or of a weaker form. In either case there is a high degree of regularity. This regularity is essential to what is called choice. When a person makes a choice, his choice is part of one of the world's causal chains. If no compulsion is involved, which means that the choice is based on his own preference, arising out of his own character, there is no reason for not calling it a free choice. It is true that his character caused him to choose as he did, and, this in turn, is conditioned by previous causes. But there is no reason for saying that his character *compelled* him to choose as he did, because the word "compel" is defined in terms of outside causal factors. Of course, it is possible for a psychotic to be in a highly abnormal mental state; it could be said that he committed a crime because his nature compelled him to do so. But the term "compel" here is used because it is felt that his abnormality prevented him from seeing clearly the consequences of various courses of action. It rendered him incapable of rational deliberation and decision. There is a serious problem here of where to draw the line between premeditated, willful behavior, and actions compelled by abnormal mental states. In general, however, free choice is a decision made by someone capable of foreseeing the consequences of alternative courses of action and choosing that which he prefers. From my point of view there is no contradiction between free choice, understood in this way, and determinism, even of the strong classical type.

In recent years, a number of writers have suggested that indeterminate quantum jumps, which most physicists believe to be random in a basic sense, may play a role in decision making.[2] Now it is quite true that under certain conditions a microcause, such as a quantum jump, can lead to an observable macroeffect. In an atom bomb, for example, a chain reaction is set off only when a sufficient number of neutrons are freed. It is also possible that in the human organism, more so than in most inanimate physical systems, there are certain points where one single quantum jump can lead to an observable macroeffect. But it is not likely that these are points at which human decisions are made.

Think for a moment of a human being at the instant of making a decision. If, at this point, there is the type of indeterminacy exhibited by a quantum jump, then the decision made at that point would be equally random. Such randomness is of no help in strengthening the meaning of the term "free choice". A choice like this would not be a

---

[2] Henry Margenau makes this point in his *Open Vistas: Philosophical Perspectives of Modern Science* (New Haven: Yale University Press, 1961). Philipp Frank, *Philosophy of Science* (Englewood, N.J.: Prentice-Hall, 1957), Chapter 10, Section 4, gives quotations from many authors on both sides of the controversy.

choice at all, but would be a chance, haphazard decision, as though a decision between two possible courses of action were made by flipping a penny. Fortunately, the range of indeterminacy in quantum theory is extremely small. If it were much greater, there might be times when a table would suddenly explode, or a falling stone would spontaneously move horizontally or back up in the air. It might be possible to survive in such a world, but surely it would not increase the possibility of free choices. On the contrary, it would make such choices considerably more difficult because it would be more difficult to anticipate the consequences of actions. When a stone is dropped, it is expected to fall to the ground. Instead, it spirals around and hits someone on the head. Then one would be thought responsible for it when there really was no intention. It is apparent then that if the consequences of actions were more difficult to foresee than they are now, the probabilities would be smaller that desired effects would take place. This would make deliberate moral behavior vastly more difficult. The same applies to random processes which may exist within the human organism. To the extent that they influence choices, they would simply be adding haphazardry to the choices. There would be *less* choice than otherwise, and an even more destructive argument could be made against the possibility of free will.

In my opinion, on the practical level of everyday life, there is no difference between classical physics, with its strong determinism, and modern quantum physics, with its random microeffects. The uncertainty in quantum theory is so very much smaller than the uncertainty in daily life arising from the limitations of knowledge. Here is a man in a world as described by classical physics. There is a man in a world as described by modern physics. There is no difference in the two descriptions that would have any significant effect on the question of free choice and moral behavior. In both cases, the man can predict the results of his actions, not with certainty, but only with some degree of probability. The indeterminacy in quantum mechanics has no observable effect on what happens to a stone when each man throws it, because the stone is an enormous complex consisting of billions of particles. In the macroworld with which human beings are concerned, the indeterminacy of quantum mechanics plays no role. For this reason I regard it as a misconception to suppose that indeterminacy on the subatomic level has any bearing on the question of free decision. However, a number of prominent scientists and philosophers of science think otherwise, and this should be accepted only as my own opinion.

Part V

*THEORETICAL
LAWS AND
THEORETICAL
CONCEPTS*

CHAPTER 23

Theories and
Nonobservables

ONE OF THE most important distinctions
between two types of laws in science is the distinction between what
may be called (there is no generally accepted terminology for them)
empirical laws and theoretical laws. Empirical laws are laws that can
be confirmed directly by empirical observations. The term "observable"
is often used for any phenomenon that can be directly observed, so it
can be said that empirical laws are laws about observables.

Here, a warning must be issued. Philosophers and scientists have
quite different ways of using the terms "observable" and "nonobservable". To a philosopher, "observable" has a very narrow meaning. It
applies to such properties as "blue", "hard", "hot". These are properties directly perceived by the senses. To the physicist, the word has a
much broader meaning. It includes any quantitative magnitude that can
be measured in a relatively simple, direct way. A philosopher would not
consider a temperature of, perhaps, 80 degrees centigrade, or a weight
of 93½ pounds, an observable because there is no direct sensory perception of such magnitudes. To a physicist, both are observables because they can be measured in an extremely simple way. The object to

be weighed is placed on a balance scale. The temperature is measured with a thermometer. The physicist would not say that the mass of a molecule, let alone the mass of an electron, is something observable, because here the procedures of measurement are much more complicated and indirect. But magnitudes that can be established by relatively simple procedures—length with a ruler, time with a clock, or frequency of light waves with a spectrometer—are called observables.

A philosopher might object that the intensity of an electric current is not really observed. Only a pointer position was observed. An ammeter was attached to the circuit and it was noted that the pointer pointed to a mark labeled 5.3. Certainly the current's intensity was not observed. It was *inferred* from what was observed.

The physicist would reply that this was true enough, but the inference was not very complicated. The procedure of measurement is so simple, so well established, that it could not be doubted that the ammeter would give an accurate measurement of current intensity. Therefore, it is included among what are called observables.

There is no question here of who is using the term "observable" in a right or proper way. There is a continuum which starts with direct sensory observations and proceeds to enormously complex, indirect methods of observation. Obviously no sharp line can be drawn across this continuum; it is a matter of degree. A philosopher is sure that the sound of his wife's voice, coming from across the room, is an observable. But suppose he listens to her on the telephone. Is her voice an observable or isn't it? A physicist would certainly say that when he looks at something through an ordinary microscope, he is observing it directly. Is this also the case when he looks into an electron microscope? Does he observe the path of a particle when he sees the track it makes in a bubble chamber? In general, the physicist speaks of observables in a very wide sense compared with the narrow sense of the philosopher, but, in both cases, the line separating observable from nonobservable is highly arbitrary. It is well to keep this in mind whenever these terms are encountered in a book by a philosopher or scientist. Individual authors will draw the line where it is most convenient, depending on their points of view, and there is no reason why they should not have this privilege.

Empirical laws, in my terminology, are laws containing terms either directly observable by the senses or measurable by relatively simple techniques. Sometimes such laws are called empirical generalizations, as a reminder that they have been obtained by generalizing results found by observations and measurements. They include not only simple qualitative laws (such as, "All ravens are black") but also quantitative laws that arise from simple measurements. The laws relating pressure, volume, and temperature of gases are of this type. Ohm's law, connecting the electric potential difference, resistance, and intensity of current, is another familiar example. The scientist makes repeated measurements, finds certain regularities, and expresses them in a law. These are the empirical laws. As indicated in earlier chapters, they are used for explaining observed facts and for predicting future observable events.

There is no commonly accepted term for the second kind of laws, which I call *theoretical laws*. Sometimes they are called abstract or hypothetical laws. "Hypothetical" is perhaps not suitable because it suggests that the distinction between the two types of laws is based on the degree to which the laws are confirmed. But an empirical law, if it is a tentative hypothesis, confirmed only to a low degree, would still be an empirical law although it might be said that it was rather hypothetical. A theoretical law is not to be distinguished from an empirical law by the fact that it is not well established, but by the fact that it contains terms of a different kind. The terms of a theoretical law do not refer to observables even when the physicist's wide meaning for what can be observed is adopted. They are laws about such entities as molecules, atoms, electrons, protons, electromagnetic fields, and others that cannot be measured in simple, direct ways.

If there is a static field of large dimensions, which does not vary from point to point, physicists call it an observable field because it can be measured with a simple apparatus. But if the field changes from point to point in very small distances, or varies very quickly in time, perhaps changing billions of times each second, then it cannot be directly measured by simple techniques. Physicists would not call such a field an observable. Sometimes a physicist will distinguish between observables and nonobservables in just this way. If the magnitude remains the same within large enough spatial distances, or large enough time intervals, so that an apparatus can be applied for a direct measurement of the magnitude, it is called a *macroevent*. If the magnitude changes within such extremely small intervals of space and time that it cannot be directly measured by simple apparatus, it is a *microevent*. (Earlier authors used the terms "microscopic" and "macroscopic", but today many authors have shortened these terms to "micro" and "macro".)

A microprocess is simply a process involving extremely small intervals of space and time. For example, the oscillation of an electromagnetic wave of visible light is a microprocess. No instrument can directly measure how its intensity varies. The distinction between macro- and microconcepts is sometimes taken to be parallel to observable and nonobservable. It is not exactly the same, but it is roughly so. Theoretical laws concern nonobservables, and very often these are microprocesses. If so, the laws are sometimes called microlaws. I use the term "theoretical laws" in a wider sense than this, to include all those laws that contain nonobservables, regardless of whether they are microconcepts or macroconcepts.

It is true, as shown earlier, that the concepts "observable" and "nonobservable" cannot be sharply defined because they lie on a continuum. In actual practice, however, the difference is usually great enough so there is not likely to be debate. All physicists would agree that the laws relating pressure, volume, and temperature of a gas, for example, are empirical laws. Here the amount of gas is large enough so that the magnitudes to be measured remain constant over a sufficiently large volume of space and period of time to permit direct, simple measurements which can then be generalized into laws. All physicists would agree that laws about the behavior of single molecules are theoretical. Such laws concern a microprocess about which generalizations cannot be based on simple, direct measurements.

Theoretical laws are, of course, more general than empirical laws. It is important to understand, however, that theoretical laws cannot be arrived at simply by taking the empirical laws, then generalizing a few steps further. How does a physicist arrive at an empirical law? He observes certain events in nature. He notices a certain regularity. He describes this regularity by making an inductive generalization. It might be supposed that he could now put together a group of empirical laws, observe some sort of pattern, make a wider inductive generalization, and arrive at a theoretical law. Such is not the case.

To make this clear, suppose it has been observed that a certain iron bar expands when heated. After the experiment has been repeated many times, always with the same result, the regularity is generalized by saying that this bar expands when heated. An empirical law has been stated, even though it has a narrow range and applies only to one particular iron bar. Now further tests are made of other iron objects with the ensuing discovery that every time an iron object is heated it expands. This permits a more general law to be formulated, namely that all bodies of iron expand when heated. In similar fashion, the still more general laws "All metals . . .", then "All solid bodies . . .", are developed. These are all simple generalizations, each a bit more general than the previous one, but they are all empirical laws. Why? Because in each case, the objects dealt with are observable (iron, copper, metal, solid bodies); in each case the increases in temperature and length are measurable by simple, direct techniques.

In contrast, a theoretical law relating to this process would refer to the behavior of molecules in the iron bar. In what way is the behavior of the molecules connected with the expansion of the bar when heated? You see at once that we are now speaking of nonobservables. We must introduce a theory—the atomic theory of matter—and we are quickly plunged into atomic laws involving concepts radically different from those we had before. It is true that these theoretical concepts differ from concepts of length and temperature only in the degree to which they are directly or indirectly observable, but the difference is so great that there is no debate about the radically different nature of the laws that must be formulated.

Theoretical laws are related to empirical laws in a way somewhat analogous to the way empirical laws are related to single facts. An empirical law helps to explain a fact that has been observed and to predict a fact not yet observed. In similar fashion, the theoretical law helps to explain empirical laws already formulated, and to permit the derivation of new empirical laws. Just as the single, separate facts fall into place in an orderly pattern when they are generalized in an empirical law, the single and separate empirical laws fit into the orderly pattern of a theoretical law. This raises one of the main problems in the methodology of science. How can the kind of knowledge that will justify the assertion of a theoretical law be obtained? An empirical law may be justified by making observations of single facts. But to justify a theoretical law, comparable observations cannot be made because the entities referred to in theoretical laws are nonobservables.

Before taking up this problem, some remarks made in an earlier chapter, about the use of the word "fact", should be repeated. It is important in the present context to be extremely careful in the use of this word because some authors, especially scientists, use "fact" or "empirical fact" for some propositions which I would call empirical laws. For example, many physicists will refer to the "fact" that the specific heat of copper is .090. I would call this a law because in its full formulation it is seen to be a universal conditional statement: "For any x, and any time t, if x is a solid body of copper, then the specific heat of x at t is .090." Some physicists may even speak of the law of thermal expansion, Ohm's law, and others, as facts. Of course, they can then say that theoretical laws help explain such facts. This sounds like my statement that empirical laws explain facts, but the word "fact" is being used here in two different ways. I restrict the word to particular, concrete facts that can be spatiotemporally specified, not thermal expansion in general, but *the* expansion of this iron bar observed this morning at ten o'clock when it was heated. It is important to bear in mind the restricted way in which I speak of facts. If the word "fact" is used in an ambiguous manner, the important difference between the ways in which empirical and theoretical laws serve for explanation will be entirely blurred.

How can theoretical laws be discovered? We cannot say: "Let's just collect more and more data, then generalize beyond the empirical laws until we reach theoretical ones." No theoretical law was ever found that way. We observe stones and trees and flowers, noting various regularities and describing them by empirical laws. But no matter how long or how carefully we observe such things, we never reach a point at which we observe a molecule. The term "molecule" never arises as a result of observations. For this reason, no amount of generalization from observations will ever produce a theory of molecular processes. Such a theory must arise in another way. It is stated not as a generalization of facts but as a hypothesis. The hypothesis is then tested in a manner analogous in certain ways to the testing of an empirical law. From the hypothesis, certain empirical laws are derived, and these empirical laws are tested in turn by observation of facts. Perhaps the empirical laws derived from the theory are already known and well confirmed. (Such laws may even have motivated the formulation of the theoretical law.) Regardless of whether the derived empirical laws are known and confirmed, or whether they are new laws confirmed by new observations, the confirmation of such derived laws provides indirect confirmation of the theoretical law.

The point to be made clear is this. A scientist does not start with one empirical law, perhaps Boyle's law for gases, and then seek a theory about molecules from which this law can be derived. The scientist tries to formulate a much more general theory from which a variety of empirical laws can be derived. The more such laws, the greater their variety and apparent lack of connection with one another, the stronger will be the theory that explains them. Some of these derived laws may have been known before, but the theory may also make it possible to derive new empirical laws which can be confirmed by new tests. If this is the case, it can be said that the theory made it possible to predict new empirical laws. The prediction is understood in a hypothetical way. If the theory holds, certain empirical laws will also hold. The predicted empirical law speaks about relations between observables, so it is now possible to make experiments to see if the empirical law holds. If the empirical law is confirmed, it provides indirect confirmation of the theory. Every confirmation of a law, empirical or theoretical, is, of course, only partial, never complete and absolute. But in the case of empirical laws, it is a more direct confirmation. The confirmation of a theoretical law is indirect, because it takes place only through the confirmation of empirical laws derived from the theory.

The supreme value of a new theory is its power to predict new empirical laws. It is true that it also has value in explaining known empirical laws, but this is a minor value. If a scientist proposes a new theoretical system, from which no new laws can be derived, then it is logically equivalent to the set of all known empirical laws. The theory may have a certain elegance, and it may simplify to some degree the set of all known laws, although it is not likely that there would be an essential simplification. On the other hand, every new theory in physics that has led to a great leap forward has been a theory from which new empirical laws could be derived. If Einstein had done no more than propose his theory of relativity as an elegant new theory that would embrace certain known laws—perhaps also simplify them to a certain degree—then his theory would not have had such a revolutionary effect.

Of course it was quite otherwise. The theory of relativity led to new empirical laws which explained for the first time such phenomena as the movement of the perihelion of Mercury, and the bending of light rays in the neighborhood of the sun. These predictions showed that relativity theory was more than just a new way of expressing the old laws. Indeed, it was a theory of great predictive power. The consequences that can be derived from Einstein's theory are far from being exhausted. These are consequences that could not have been derived from earlier theories. Usually a theory of such power does have an elegance, and a unifying effect on known laws. It is simpler than the total collection of known laws. But the great value of the theory lies in its power to suggest new laws that can be confirmed by empirical means.

CHAPTER 24

Correspondence Rules

AN IMPORTANT qualification must now be added to the discussion of theoretical laws and terms given in the last chapter. The statement that empirical laws are derived from theoretical laws is an oversimplification. It is not possible to derive them directly because a theoretical law contains theoretical terms, whereas an empirical law contains only observable terms. This prevents any direct deduction of an empirical law from a theoretical one.

To understand this, imagine that we are back in the nineteenth century, preparing to state for the first time some theoretical laws about molecules in a gas. These laws are to describe the number of molecules per unit volume of the gas, the molecular velocities, and so forth. To simplify matters, we assume that all the molecules have the same velocity. (This was indeed the original assumption; later it was abandoned in favor of a certain probability distribution of velocities.) Further assumptions must be made about what happens when molecules collide. We do not know the exact shape of molecules, so let us suppose that they are tiny spheres. How do spheres collide? There are laws about colliding spheres, but they concern large bodies. Since we cannot

232

directly observe molecules, we assume their collisions are analogous to those of large bodies; perhaps they behave like perfect billiard balls on a frictionless table. These are, of course, only assumptions; guesses suggested by analogies with known macrolaws.

But now we come up against a difficult problem. Our theoretical laws deal exclusively with the behavior of molecules, which cannot be seen. How, therefore, can we deduce from such laws a law about observable properties such as the pressure or temperature of a gas or properties of sound waves that pass through the gas? The theoretical laws contain only theoretical terms. What we seek are empirical laws containing observable terms. Obviously, such laws cannot be derived without having something else given in addition to the theoretical laws.

The something else that must be given is this: a set of rules connecting the theoretical terms with the observable terms. Scientists and philosophers of science have long recognized the need for such a set of rules, and their nature has been often discussed. An example of such a rule is: "If there is an electromagnetic oscillation of a specified frequency, then there is a visible greenish-blue color of a certain hue." Here something observable is connected with a nonobservable microprocess.

Another example is: "The temperature (measured by a thermometer and, therefore, an observable in the wider sense explained earlier) of a gas is proportional to the mean kinetic energy of its molecules." This rule connects a nonobservable in molecular theory, the kinetic energy of molecules, with an observable, the temperature of the gas. If statements of this kind did not exist, there would be no way of deriving empirical laws about observables from theoretical laws about nonobservables.

Different writers have different names for these rules. I call them "correspondence rules". P. W. Bridgman calls them operational rules. Norman R. Campbell speaks of them as the "Dictionary".[1] Since the rules connect a term in one terminology with a term in another terminology, the use of the rules is analogous to the use of a French-English dictionary. What does the French word "cheval" mean? You look it up in the dictionary and find that it means "horse". It is not really that

[1] See Percy W. Bridgman, *The Logic of Modern Physics* (New York: Macmillan, 1927), and Norman R. Campbell, *Physics: The Elements* (Cambridge: Cambridge University Press, 1920); reprinted as *Foundations of Science* (New York: Dover, 1957). Rules of correspondence are discussed by Ernest Nagel, *The Structure of Science* (New York: Harcourt, Brace & World, 1961), pp. 97–105.

simple when a set of rules is used for connecting nonobservables with observables; nevertheless, there is an analogy here that makes Campbell's "Dictionary" a suggestive name for the set of rules.

There is a temptation at times to think that the set of rules provides a means for defining theoretical terms, whereas just the opposite is really true. A theoretical term can never be explicitly defined on the basis of observable terms, although sometimes an observable can be defined in theoretical terms. For example, "iron" can be defined as a substance consisting of small crystalline parts, each having a certain arrangement of atoms and each atom being a configuration of particles of a certain type. In theoretical terms then, it is possible to express what is meant by the observable term "iron", but the reverse is not true.

There is no answer to the question: "Exactly what is an electron?" Later we shall come back to this question, because it is the kind that philosophers are always asking scientists. They want the physicist to tell them just what he means by "electricity", "magnetism", "gravity", "a molecule". If the physicist explains them in theoretical terms, the philosopher may be disappointed. "That is not what I meant at all", he will say. "I want you to tell me, in ordinary language, what those terms mean." Sometimes the philosopher writes a book in which he talks about the great mysteries of nature. "No one", he writes, "has been able so far, and perhaps no one ever will be able, to give us a straightforward answer to the question: 'What is electricity?' And so electricity remains forever one of the great, unfathomable mysteries of the universe."

There is no special mystery here. There is only an improperly phrased question. Definitions that cannot, in the nature of the case, be given, should not be demanded. If a child does not know what an elephant is, we can tell him it is a huge animal with big ears and a long trunk. We can show him a picture of an elephant. It serves admirably to define an elephant in observable terms that a child can understand. By analogy, there is a temptation to believe that, when a scientist introduces theoretical terms, he should also be able to define them in familiar terms. But this is not possible. There is no way a physicist can show us a picture of electricity in the way he can show his child a picture of an elephant. Even the cell of an organism, although it cannot be seen with the unaided eye, can be represented by a picture because the cell can be seen when it is viewed through a microscope. But we do not possess a picture of the electron. We cannot say how it looks or how it feels, because it cannot be seen or touched. The best we can do is to say that it is an extremely small body that behaves in a certain manner. This may seem to be analogous to our description of an elephant. We can describe an elephant as a large animal that behaves in a certain manner. Why not do the same with an electron?

The answer is that a physicist can describe the behavior of an electron only by stating theoretical laws, and these laws contain only theoretical terms. They describe the field produced by an electron, the reaction of an electron to a field, and so on. If an electron is in an electrostatic field, its velocity will accelerate in a certain way. Unfortunately, the electron's acceleration is an unobservable. It is not like the acceleration of a billiard ball, which can be studied by direct observation. There is no way that a theoretical concept can be defined in terms of observables. We must, therefore, resign ourselves to the fact that definitions of the kind that can be supplied for observable terms cannot be formulated for theoretical terms.

It is true that some authors, including Bridgman, have spoken of the rules as "operational definitions". Bridgman had a certain justification, because he used his rules in a somewhat different way, I believe, than most physicists use them. He was a great physicist and was certainly aware of his departure from the usual use of rules, but he was willing to accept certain forms of speech that are not customary, and this explains his departure. It was pointed out in a previous chapter that Bridgman preferred to say that there is not just one concept of intensity of electric current, but a dozen concepts. Each procedure by which a magnitude can be measured provides an operational definition for that magnitude. Since there are different procedures for measuring current, there are different concepts. For the sake of convenience, the physicist speaks of just one concept of current. Strictly speaking, Bridgman believed, he should recognize many different concepts, each defined by a different operational procedure of measurement.

We are faced here with a choice between two different physical languages. If the customary procedure among physicists is followed, the various concepts of current will be replaced by one concept. This means, however, that you place the concept in your theoretical laws, because the operational rules are just correspondence rules, as I call them, which connect the theoretical terms with the empirical ones. Any claim to possessing a definition—that is, an operational definition—of the theoretical concept must be given up. Bridgman could speak of having operational definitions for his theoretical terms only because he was not speaking of a general concept. He was speaking of partial concepts, each defined by a different empirical procedure.

Even in Bridgman's terminology, the question of whether his partial concepts can be adequately defined by operational rules is problematic. Reichenbach speaks often of what he calls "coordinating definitions". (In his German publications, he calls them *Zuordnungsdefinitionen,* from *zuordnen,* which means to coordinate.) Perhaps "coordination" is a better term than "definition" for what Bridgman's rules actually do. In geometry, for instance, Reichenbach points out that the axiom system of geometry, as developed by David Hilbert, for example, is an uninterpreted axiom system. The basic concepts of point, line, and plane could just as well be called "class alpha", "class beta", and "class gamma". We must not be seduced by the sound of familiar words, such as "point" and "line", into thinking they must be taken in their ordinary meaning. In the axiom system, they are uninterpreted terms. But when geometry is applied to physics, these terms must be connected with something in the physical world. We can say, for example, that the lines of the geometry are exemplified by rays of light in a vacuum or by stretched cords. In order to connect the uninterpreted terms with observable physical phenomena, we must have rules for establishing the connection.

What we call these rules is, of course, only a terminological question; we should be cautious and not speak of them as definitions. They are not definitions in any strict sense. We cannot give a really adequate definition of the geometrical concept of "line" by referring to anything in nature. Light rays, stretched strings, and so on are only approximately straight; moreover, they are not lines, but only segments of lines. In geometry, a line is infinite in length and absolutely straight. Neither property is exhibited by any phenomenon in nature. For that reason, it is not possible to give an operational definition, in the strict sense of the word, of concepts in theoretical geometry. The same is true of all the other theoretical concepts of physics. Strictly speaking, there are no "definitions" of such concepts. I prefer not to speak of "operational definitions" or even to use Reichenbach's term "coordinating definitions". In my publications (only in recent years have I written about this question), I have called them "rules of correspondence" or, more simply, "correspondence rules".

Campbell and other authors often speak of the entities in theoretical physics as mathematical entities. They mean by this that the entities are related to each other in ways that can be expressed by mathematical functions. But they are not mathematical entities of the sort that can be defined in pure mathematics. In pure mathematics, it is possible to define various kinds of numbers, the function of logarithm, the exponential function, and so forth. It is not possible, however, to define such terms as "electron" and "temperature" by pure mathematics. Physical terms can be introduced only with the help of nonlogical constants, based on observations of the actual world. Here we have an essential difference between an axiomatic system in mathematics and an axiomatic system in physics.

If we wish to give an interpretation to a term in a mathematical axiom system, we can do it by giving a definition in logic. Consider, for example, the term "number" as it is used in Peano's axiom system. We can define it in logical terms, by the Frege-Russell method, for example. In this way the concept of "number" acquires a complete, explicit definition on the basis of pure logic. There is no need to establish a connection between the number 5 and such observables as "blue" and "hot". The terms have only a logical interpretation; no connection with the actual world is needed. Sometimes an axiom system in mathematics is called a theory. Mathematicians speak of set theory, group theory, matrix theory, probability theory. Here the word "theory" is used in a purely analytic way. It denotes a deductive system that makes no reference to the actual world. We must always bear in mind that such a use of the word "theory" is entirely different from its use in reference to empirical theories such as relativity theory, quantum theory, psychoanalytical theory, and Keynesian economic theory.

A postulate system in physics cannot have, as mathematical theories have, a splendid isolation from the world. Its axiomatic terms—"electron", "field", and so on—must be interpreted by correspondence rules that connect the terms with observable phenomena. This interpretation is necessarily incomplete. Because it is always incomplete, the system is left open to make it possible to add new rules of correspondence. Indeed, this is what continually happens in the history of physics. I am not thinking now of a revolution in physics, in which an entirely new theory is developed, but of less radical changes that modify existing theories. Nineteenth-century physics provides a good example, because classical mechanics and electromagnetics had been established, and, for many decades, there was relatively little change in fundamental laws. The basic theories of physics remained unchanged. There was, however, a steady addition of new correspondence rules, because new procedures were continually being developed for measuring this or that magnitude.

Of course, physicists always face the danger that they may develop correspondence rules that will be incompatible with each other or with the theoretical laws. As long as such incompatibility does not occur, however, they are free to add new correspondence rules. The procedure is never-ending. There is always the possibility of adding new rules, thereby increasing the amount of interpretation specified for the theoretical terms; but no matter how much this is increased, the interpretation is never final. In a mathematical system, it is otherwise. There a logical interpretation of an axiomatic term *is* complete. Here we find another reason for reluctance in speaking of theoretical terms as "defined" by correspondence rules. It tends to blur the important distinction between the nature of an axiom system in pure mathematics and one in theoretical physics.

Is it not possible to interpret a theoretical term by correspondence rules so completely that no further interpretation would be possible? Perhaps the actual world is limited in its structure and laws. Eventually a point may be reached beyond which there will be no room for strengthening the interpretation of a term by new correspondence rules. Would not the rules then provide a final, explicit definition for the term? Yes, but then the term would no longer be theoretical. It would become part of the observation language. The history of physics has not yet indicated that physics will become complete; there has been only a steady addition of new correspondence rules and a continual modification in the interpretations of theoretical terms. There is no way of knowing whether this is an infinite process or whether it will eventually come to some sort of end.

It may be looked at this way. There is no prohibition in physics against making the correspondence rules for a term so strong that the term becomes explicitly defined and therefore ceases to be theoretical. Neither is there any basis for assuming that it will always be possible to add new correspondence rules. Because the history of physics has shown such a steady, unceasing modification of theoretical concepts, most physicists would advise against correspondence rules so strong that a theoretical term becomes explicitly defined. Moreover, it is a wholly unnecessary procedure. Nothing is gained by it. It may even have the adverse effect of blocking progress.

Of course, here again we must recognize that the distinction between observables and nonobservables is a matter of degree. We might give an explicit definition, by empirical procedures, to a concept such as length, because it is so easily and directly measured, and is unlikely to be modified by new observations. But it would be rash to seek such strong correspondence rules that "electron" would be explicitly defined. The concept "electron" is so far removed from simple, direct observations that it is best to keep it theoretical, open to modifications by new observations.

CHAPTER 25

How New Empirical Laws Are Derived from Theoretical Laws

IN CHAPTER 24, the discussion concerned the ways in which correspondence rules are used for linking the non-observable terms of a theory with the observable terms of empirical laws. This can be made clearer by a few examples of the manner in which empirical laws have actually been derived from the laws of a theory.

The first example concerns the kinetic theory of gases. Its model, or schematic picture, is one of small particles called molecules, all in constant agitation. In its original form, the theory regarded these particles as little balls, all having the same mass and, when the temperature of the gas is constant, the same constant velocity. Later it was discovered that the gas would not be in a stable state if each particle had the same velocity; it was necessary to find a certain probability distribution of velocities that would remain stable. This was called the Maxwell-Boltzmann distribution. According to this distribution, there was a certain probability that any molecule would be within a certain range on the velocity scale.

When the kinetic theory was first developed, many of the magnitudes occurring in the laws of the theory were not known. No one knew the mass of a molecule, or how many molecules a cubic centimeter of gas at a certain temperature and pressure would contain. These magnitudes were expressed by certain parameters written into the laws. After the equations were formulated, a dictionary of correspondence rules was prepared. These correspondence rules connected the theoretical terms with observable phenomena in a way that made it possible to determine indirectly the values of the parameters in the equations. This, in turn, made it possible to derive empirical laws. One correspondence rule states that the temperature of the gas corresponds to the mean kinetic energy of the molecules. Another correspondence rule connects the pressure of the gas with the impact of molecules on the confining wall of a vessel. Although this is a discontinuous process involving discrete molecules, the total effect can be regarded as a constant force pressing on the wall. Thus, by means of correspondence rules, the pressure that is measured macroscopically by a manometer (pressure gauge) can be expressed in terms of the statistical mechanics of molecules.

What is the density of the gas? Density is mass per unit volume, but how do we measure the mass of a molecule? Again our dictionary— a very simple dictionary—supplies the correspondence rule. The total mass M of the gas is the sum of the masses m of the molecules. M is observable (we simply weigh the gas), but m is theoretical. The dictionary of correspondence rules gives the connection between the two concepts. With the aid of this dictionary, empirical tests of various laws derived from our theory are possible. On the basis of the theory, it is possible to calculate what will happen to the pressure of the gas when its volume remains constant and its temperature is increased. We can calculate what will happen to a sound wave produced by striking the side of the vessel, and what will happen if only part of the gas is heated. These theoretical laws are worked out in terms of various parameters that occur within the equations of the theory. The dictionary of correspondence rules enables us to express these equations as empirical laws, in which concepts are measurable, so that empirical procedures can supply values for the parameters. If the empirical laws can be confirmed, this provides indirect confirmation of the theory. Many of the empirical laws for gases were known, of course, before the kinetic theory was developed. For these laws, the theory provided an explanation. In addition, the theory led to previously unknown empirical laws.

The power of a theory to predict new empirical laws is strikingly exemplified by the theory of electromagnetism, which was developed about 1860 by two great English physicists, Michael Faraday and James Clerk Maxwell. (Faraday did most of the experimental work, and Maxwell did most of the mathematical work.) The theory dealt with electric charges and how they behaved in electrical and magnetic fields. The concept of the electron—a tiny particle with an elementary electric charge—was not formulated until the very end of the century. Maxwell's famous set of differential equations, for describing electromagnetic fields, presupposed only small discrete bodies of unknown nature, capable of carrying an electric charge or a magnetic pole. What happens when a current moves along a copper wire? The theory's dictionary made this observable phenomenon correspond to the actual movement along the wire of little charged bodies. From Maxwell's theoretical model, it became possible (with the help of correspondence rules, of course) to derive many of the known laws of electricity and magnetism.

The model did much more than this. There was a certain parameter c in Maxwell's equations. According to his model, a disturbance in an electromagnetic field would be propagated by waves having the velocity c. Electrical experiments showed the value of c to be approximately 3×10^{10} centimeters per second. This was the same as the known value for the speed of light, and it seemed unlikely that it was an accident. Is it possible, physicists asked themselves, that light is simply a special case of the propagation of an electromagnetic oscillation? It was not long before Maxwell's equations were providing explanations for all sorts of optical laws, including refraction, the velocity of light in different media, and many others.

Physicists would have been pleased enough to find that Maxwell's model explained known electrical and magnetic laws; but they received a double bounty. The theory also explained optical laws! Finally, the great strength of the new model was revealed in its power to predict, to formulate empirical laws that had not been previously known.

The first instance was provided by Heinrich Hertz, the German physicist. About 1890, he began his famous experiments to see whether electromagnetic waves of low frequency could be produced and detected in the laboratory. Light is an electromagnetic oscillation and propagation of waves at very high frequency. But Maxwell's laws made it possible for such waves to have *any* frequency. Hertz's experiments resulted in his discovery of what at first were called Hertz waves. They are now called radio waves. At first, Hertz was able to transmit these waves from one oscillator to another over only a small distance—first a few centimeters, then a meter or more. Today a radio broadcasting station sends its waves many thousands of miles.

The discovery of radio waves was only the beginning of the derivation of new laws from Maxwell's theoretical model. X rays were discovered and were thought at first to be particles of enormous velocity and penetrative power. Then it occurred to physicists that, like light and radio waves, these might be electromagnetic waves, but of extremely high frequency, much higher than the frequency of visible light. This also was later confirmed, and laws dealing with X rays were derived from Maxwell's fundamental field equations. X rays proved to be waves of a certain frequency range within the much broader frequency band of gamma rays. The X rays used today in medicine are simply gamma rays of certain frequency. All this was largely predictable on the basis of Maxwell's model. His theoretical laws, together with the correspondence rules, led to an enormous variety of new empirical laws.

The great variety of fields in which experimental confirmation was found contributed especially to the strong overall confirmation of Maxwell's theory. The various branches of physics had originally developed for practical reasons; in most cases, the divisions were based on our different sense organs. Because the eyes perceive light and color, we call such phenomena optics; because our ears hear sounds, we call a branch of physics acoustics; and because our bodies feel heat, we have a theory of heat. We find it useful to construct simple machines based on the movements of bodies, and we call it mechanics. Other phenomena, such as electricity and magnetism, cannot be directly perceived, but their consequences can be observed.

In the history of physics, it is always a big step forward when one branch of physics can be explained by another. Acoustics, for instance, was found to be only a part of mechanics, because sound waves are simply elasticity waves in solids, liquids, and gases. We have already spoken of how the laws of gases were explained by the mechanics of moving molecules. Maxwell's theory was another great leap forward toward the unification of physics. Optics was found to be a part of electromagnetic theory. Slowly the notion grew that the whole of physics might some day be unified by one great theory. At present there is an enormous gap between electromagnetism on the one side and gravitation on the other. Einstein made several attempts to develop a unified field theory that might close this gap; more recently, Heisenberg and others have made similar attempts. So far, however, no theory has been devised that is entirely satisfactory or that provides new empirical laws capable of being confirmed.

Physics originally began as a descriptive macrophysics, containing an enormous number of empirical laws with no apparent connections. In the beginning of a science, scientists may be very proud to have discovered hundreds of laws. But, as the laws proliferate, they become unhappy with this state of affairs; they begin to search for underlying, unifying principles. In the nineteenth century, there was considerable controversy over the question of underlying principles. Some felt that science must find such principles, because otherwise it would be no more than a description of nature, not a real explanation. Others thought that that was the wrong approach, that underlying principles belong only to metaphysics. They felt that the scientist's task is merely to describe, to find out *how* natural phenomena occur, not *why*.

Today we smile a bit about the great controversy over description versus explanation. We can see that there was something to be said for both sides, but that their way of debating the question was futile. There is no real opposition between explanation and description. Of course, if description is taken in the narrowest sense, as merely describing what a certain scientist did on a certain day with certain materials, then the opponents of mere description were quite right in asking for more, for a real explanation. But today we see that description in the broader sense, that of placing phenomena in the context of more general laws, provides the only type of explanation that can be given for phenomena. Similarly, if the proponents of explanation mean a metaphysical explanation, not grounded in empirical procedures, then their opponents were correct in insisting that science should be concerned only with description. Each side had a valid point. Both description and explanation, rightly understood, are essential aspects of science.

The first efforts at explanation, those of the Ionian natural philosophers, were certainly partly metaphysical; the world is all fire, or all water, or all change. Those early efforts at scientific explanation can be viewed in two different ways. We can say: "This is not science, but pure metaphysics. There is no possibility of confirmation, no correspondence rules for connecting the theory with observable phenomena." On the other hand, we can say: "These Ionian theories are certainly not scientific, but at least they are pictorial visions of theories. They are the first primitive beginnings of science."

It must not be forgotten that, both in the history of science and in the psychological history of a creative scientist, a theory has often first appeared as a kind of visualization, a vision that comes as an inspiration to a scientist long before he has discovered correspondence rules that may help in confirming his theory. When Democritus said that everything consists of atoms, he certainly had not the slightest confirmation for this theory. Nevertheless, it was a stroke of genius, a profound insight, because two thousand years later his vision was confirmed. We should not, therefore, reject too rashly any anticipatory vision of a theory, provided it is one that may be tested at some future time. We are on solid ground, however, if we issue the warning that no hypothesis can claim to be scientific unless there is the *possibility* that it can be tested. It does not have to be confirmed to be a hypothesis, but there must be correspondence rules that will permit, in principle, a means of confirming or disconfirming the theory. It may be enormously difficult to think of experiments that can test the theory; this is the case today with various unified field theories that have been proposed. But if such tests are possible in principle, the theory can be called a scientific one. When a theory is first proposed, we should not demand more than this.

The development of science from early philosophy was a gradual, step-by-step process. The Ionian philosophers had only the most primitive theories. In contrast, the thinking of Aristotle was much clearer and on more solid scientific ground. He made experiments, and he knew the importance of experiments, although in other respects he was an apriorist. This was the beginning of science. But it was not until the time of Galileo Galilei, about 1600, that a really great emphasis was placed on the experimental method in preference to aprioristic reasoning about nature. Even though many of Galileo's concepts had previously been stated as theoretical concepts, he was the first to place theoretical physics on a solid empirical foundation. Certainly Newton's physics (about 1670) exhibits the first comprehensive, systematic theory, containing unobservables as theoretical concepts: the universal force of gravitation, a general concept of mass, theoretical properties of light rays, and so on. His theory of gravity was one of great generality. Between any two particles, small or large, there is a force inversely proportional to the square of the distance between them. Before Newton advanced this theory, science provided no explanation that applied to both the fall of a stone and the movements of planets around the sun.

It is very easy for us today to remark how strange it was that it never occurred to anyone before Newton that the same force might cause the apple to drop and the moon to go around the earth. In fact, this was not a thought likely to occur to anyone. It is not that the *answer* was so difficult to give; it is that nobody had asked the *question*. This is a vital point. No one had asked: "What is the relation between the forces that heavenly bodies exert upon each other and terrestrial forces that cause objects to fall to the ground?" Even to speak in such terms as "terrestrial" and "heavenly" is to make a bipartition, to cut nature into two fundamentally different regions. It was Newton's great insight to break away from this division, to assert that there is no such fundamental cleavage. There is one nature, one world. Newton's universal law of gravitation was the theoretical law that explained for the first time both the fall of an apple and Kepler's laws for the movements of planets. In Newton's day, it was a psychologically difficult, extremely daring adventure to think in such general terms.

Later, of course, by means of correspondence rules, scientists discovered how to determine the masses of astronomical bodies. Newton's theory also said that two apples, side by side on a table, attract each other. They do not move toward each other because the attracting force is extremely small and the friction on the table very large. Physicists eventually succeeded in actually measuring the gravitational forces between two bodies in the laboratory. They used a torsion balance consisting of a bar with a metal ball on each end, suspended at its center by a long wire attached to a high ceiling. (The longer and thinner the wire, the more easily the bar would turn.) Actually, the bar never came to an absolute rest but always oscillated a bit. But the mean point of the bar's oscillation could be established. After the exact position of the mean point was determined, a large pile of lead bricks was constructed near the bar. (Lead was used because of its great specific gravity. Gold has an even higher specific gravity, but gold bricks are expensive.) It was found that the mean of the oscillating bar had shifted a tiny amount to bring one of the balls on the end of the bar nearer to the lead pile. The shift was only a fraction of a millimeter, but it was enough to provide the first observation of a gravitational effect between two bodies in a laboratory—an effect that had been predicted by Newton's theory of gravitation.

It had been known before Newton that apples fall to the ground and that the moon moves around the earth. Nobody before Newton could have predicted the outcome of the experiment with the torsion balance. It is a classic instance of the power of a theory to predict a new phenomenon not previously observed.

CHAPTER 26

The Ramsey Sentence

SCIENTIFIC THEORY, in the sense in which we are using the term—theoretical postulates combined with correspondence rules that join theoretical and observational terms—has in recent years been intensely analyzed and discussed by philosophers of science. Much of this discussion is so new that it has not yet been published. In this chapter, we will introduce an important new approach to the topic, one that goes back to a little known paper by the Cambridge logician and economist, Frank Plumpton Ramsey.

Ramsey died in 1930 at the age of twenty-six. He did not live to complete a book, but after his death a collection of his papers was edited by Richard Bevan Braithwaite and published in 1931 as *The Foundations of Mathematics*.[1] A short paper entitled "Theories" appears in this book. In my opinion, this paper deserves much more recognition than it has received. Perhaps the book's title attracted only readers interested in the logical foundations of mathematics, so that other

[1] Ramsey, *The Foundations of Mathematics* (London: Routledge and Kegan Paul, 1931), reprinted in paperback, Littlefield, Adams (1960).

important papers in the book, such as the paper on theories, tended to be overlooked.

Ramsey was puzzled by the fact that the theoretical terms—terms for the objects, properties, forces, and events described in a theory—are not meaningful in the same way that observational terms—"iron rod", "hot", and "red"—are meaningful. How, then, does a theoretical term acquire meaning? Everyone agrees that it derives its meaning from the context of the theory. "Gene" derives its meaning from genetic theory. "Electron" is interpreted by the postulates of particle physics. But we are faced with many confusing, disturbing questions. How can the *empirical* meaning of a theoretical term be determined? What does a given theory tell us about the actual world? Does it describe the structure of the real world, or is it just an abstract, artificial device for bringing order into the large mass of experiences in somewhat the same way that a system of accounting makes it possible to keep orderly records of a firm's financial dealings? Can it be said that an electron "exists" in the same sense that an iron rod exists?

There are procedures that measure a rod's properties in a simple, direct manner. Its volume and weight can be determined with great accuracy. We can measure the wave lengths of light emitted by the surface of a heated iron rod and precisely define what we mean when we say that the iron rod is "red". But when we deal with the properties of theoretical entities, such as the "spin" of an elementary particle, there are only complicated, indirect procedures for giving the term an empirical meaning. First we must introduce "spin" in the context of an elaborate theory of quantum mechanics, and then the theory must be connected with laboratory observables by another complex set of postulates—the correspondence rules. Clearly, spin is not empirically grounded in the simple, direct manner that the redness of a heated iron rod is grounded. Exactly what *is* its cognitive status? How can theoretical terms, which must in some way be connected with the actual world and subject to empirical testing, be distinguished from those metaphysical terms so often encountered in traditional philosophy—terms that have no empirical meaning? How can the right of a scientist to speak of theoretical concepts be justified, without at the same time justifying the right of a philosopher to use metaphysical terms?

In seeking answers to these puzzling questions, Ramsey made a novel, startling suggestion. He proposed that the combined system of theoretical and correspondence postulates of a theory be replaced by what is today called the "Ramsey sentence of the theory". In the Ramsey sentence, which is equivalent to the theory's postulates, theoretical terms do not occur at all. In other words, the puzzling questions are neatly side-stepped by the elimination of the very terms about which the questions are raised.

Suppose we are concerned with a theory containing n theoretical terms: "T_1", "T_2", "T_3" . . . "T_n". These terms are introduced by the postulates of the theory. They are connected with directly observable terms by the theory's correspondence rules. In these correspondence rules occur m observational terms: "O_1", "O_2", "O_3" . . . "O_m". The theory itself is a conjunction of all the theoretical postulates together with all the correspondence postulates. A full statement of the theory, therefore, will contain the combined sets of T- and O-terms: "T_1", "T_2", . . . , "T_n"; "O_1", "O_2", . . . , "O_m". Ramsey proposed that, in this sentence, the full statement of the theory, all the theoretical terms are to be replaced by corresponding variables: "U_1", "U_2", . . . , "U_n", and that what logicians call "existential quantifiers"—'($\exists U_1$)', '($\exists U_2$)', . . . , '($\exists U_n$)'—be added to this formula. It is this new sentence, with its U-variables and their existential quantifiers, that is called the "Ramsey sentence".

To see exactly how this develops, consider the following example. Take the symbol "Mol" for the class of molecules. Instead of calling something "a molecule", call it "an element of Mol". Similarly, "Hymol" stands for "the class of hydrogen molecules", and "a hydrogen molecule" is "an element of Hymol". It is assumed that a space-time coordinate system has been fixed, so that a space-time point can be represented by its four coordinates: x, y, z, t. Adopt the symbol "Temp" for the concept of temperature. Then, "the (absolute) temperature of the body b, at time t, is 500" can be written, "$Temp(b,t) = 500$". Temperature is thus expressed as a relation involving a body, a time point, and a number. "The pressure of a body b, at time t", can be written, "$Press(b,t)$". The concept of mass is represented by the symbol "Mass". For "the mass of the body b (in grams) is 150" write, "$Mass(b) = 150$". Mass is a relation between a body and a number. Let "Vel" stand for the velocity of a body (it may be a macro- or a microbody). For example, "$Vel(b,t) = (r_1, r_2, r_3)$", where the right side of the equation refers to a triple of real numbers, namely, the components of the velocity in the directions of $x, y,$ and z. Vel is thus a relation concerning a body, a time coordinate, and a triple of real numbers.

Generally speaking, the theoretical language contains "class terms" (such as terms for macrobodies, microbodies, and events) and "relation terms" (such as terms for various physical magnitudes).

Consider theory *TC*. (The "*T*" stands for the theoretical postulates of the theory, and "*C*" stands for the postulates that give the correspondence rules.) The postulates of this theory include some laws from the kinetic theory of gases, laws concerning the motions of molecules, their velocities, cóllisions, and so on. There are general laws about any gas, and there are special laws about hydrogen. In addition, there are macro-gas-theory laws about the temperature, pressure, and total mass of a (macro-) gas body. Suppose that the theoretical postulates of theory *TC* contain all the terms mentioned above. For the sake of brevity, instead of writing out in full all the T-postulates, write only the theoretical terms, and indicate the connecting symbolism by dots:

(*T*)   . . . Mol . . . Hymol . . . Temp . . . Press . . . Mass . . . Vel . . .

To complete the symbolization of theory *TC*, the correspondence postulates for some, but not necessarily all, of the theoretical terms must be considered. These C-postulates may be operational rules for the measurement of temperature and pressure (that is, a description of the construction of a thermometer and a manometer and rules for determining the values of temperature and pressure from the numbers read on the scales of the instruments). The C-postulates will contain the theoretical terms "Temp" and "Press" as well as a number of observational terms: "O_1", "O_2", . . . , "O_m". Thus, the C-postulates can be expressed in a brief, abbreviated way by writing:

(*C*)   . . . Temp . . . O_1 . . . O_2 . . . O_3 . . . Press . . . O_4 . . . O_m . . .

The entire theory can now be indicated in the following form:

(*TC*)   . . . Mol . . . Hymol . . . Temp . . . Press . . . Mass . . . Vel . . . ; . . . Temp . . . O_1 . . . O_2 . . . O_3 . . . Press . . . O_4 . . . O_m . . .

To transform this theory *TC* into its Ramsey sentence, two steps are required. First, replace all the theoretical terms (class terms and relation terms) with arbitrarily chosen class and relation variables. Wherever "Mol" occurs in the theory, substitute the variable "C_1", for example. Wherever "Hymol" occurs in the theory replace it by another class variable, such as "C_2". The relation term "Temp" is replaced everywhere (both in the T and C portions of the theory) by a relation variable, such as "R_1". In the same way, "Press", "Mass", and "Vel" are replaced by three other relation variables, "R_2", "R_3", and "R_4" respectively, for example. The final result may be indicated in this way:

$$\ldots C_1 \ldots C_2 \ldots R_1 \ldots R_2 \ldots R_3 \ldots R_4 \ldots ;$$
$$\ldots R_1 \ldots O_1 \ldots O_2 \ldots O_3 \ldots R_2 \ldots$$
$$O_4 \ldots O_m \ldots$$

This result (which should be thought of as completely written out, rather than abbreviated as it is here with the help of dots) is no longer a sentence (as T, C, and TC are). It is an open sentence formula or, as it is sometimes called, a sentence form or a sentence function.

The second step, which transforms the open sentence formula into the Ramsey sentence, ^RTC, consists of writing in front of the sentence formula six existential quantifiers, one for each of the six variables:

$$(^RTC) \quad (\exists\, C_1)\, (\exists\, C_2)\, (\exists\, R_1)\, (\exists\, R_2)\, (\exists\, R_3)\, (\exists\, R_4)\, [\ldots C_1$$
$$\ldots C_2 \ldots R_1 \ldots R_2 \ldots R_3 \ldots R_4 \ldots ;$$
$$\ldots R_1 \ldots O_1 \ldots O_2 \ldots O_3 \ldots R_2 \ldots$$
$$O_4 \ldots O_m \ldots]$$

A formula preceded by an existential quantifier asserts that there is at least one entity (of the type to which it refers) that satisfies the condition expressed by the formula. Thus, the Ramsey sentence indicated above says (roughly speaking) that there is (at least) one class C_1, one class C_2, one relation R_1, one R_2, one R_3, and one R_4 such that:

(1) these six classes and relations are connected with one another in a specified way (namely, as specified in the first or T part of the formula),

(2) the two relations, R_1 and R_2, are connected with the m observational entities, $O_1, \ldots , O_m$, in a certain way (namely, as specified in the second or C part of the formula).

The important thing to note is that in the Ramsey sentence the theoretical terms have disappeared. In their place are variables. The variable "C_1" does not refer to any particular class. The assertion is only that there is at least one class that satisfies certain conditions. The meaning of the Ramsey sentence is not changed in any way if the variables are arbitrarily changed. For example, the symbols "C_1" and "C_2" can be interchanged or replaced with other arbitrary variables, such as "X_1" and "X_2". The meaning of the sentence remains the same.

It may appear that the Ramsey sentence is no more than just another somewhat roundabout way of expressing the original theory. In a sense, this is true. It is easy to show that any statement about the real world that does not contain theoretical terms—that is, any statement capable of empirical confirmation—that follows from the theory will also follow from the Ramsey sentence. In other words, the Ramsey sentence has precisely the same *explanatory and predictive power* as the original system of postulates. Ramsey was the first to see this. It was an important insight, although few of his colleagues gave it much attention. One of the exceptions was Braithwaite, who was Ramsey's friend and who edited his papers. In his book, *Scientific Explanation* (1953), Braithwaite discusses Ramsey's insight, emphasizing its importance.

The important fact is that we can now avoid all the troublesome metaphysical questions that plague the original formulation of theories and can introduce a simplification into the formulation of theories. Before, we had theoretical terms, such as "electron", of dubious "reality" because they were so far removed from the observable world. Whatever partial empirical meaning could be given to these terms could be given only by the indirect procedure of stating a system of theoretical postulates and connecting those postulates with empirical observations by means of correspondence rules. In Ramsey's way of talking about the external world, a term such as "electron" vanishes. This does not in any way imply that electrons vanish, or, more precisely, that whatever it is in the external world that is symbolized by the word "electron" vanishes. The Ramsey sentence continues to assert, through its existential quantifiers, that there is something in the external world that has all those properties that physicists assign to the electron. It does not question the existence—the "reality"—of this something. It merely proposes a different way of talking about that something. The troublesome question it avoids is not, "Do electrons exist?" but, "What is the exact *meaning* of the term 'electron'?" In Ramsey's way of speaking about the world, this question does not arise. It is no longer necessary to inquire about the meaning of "electron", because the term itself does not appear in Ramsey's language.

It is important to understand—and this point was not sufficiently stressed by Ramsey—that Ramsey's approach cannot be said to bring theories into the observation language if "observation language" means (as is often the case) a language containing only observational terms and the terms of elementary logic and mathematics. Modern physics demands extremely complicated, high-level mathematics. Relativity theory, for instance, calls for non-Euclidean geometry and tensor calculus, and quantum mechanics calls for equally sophisticated mathematical concepts. It cannot be said, therefore, that a physical theory, expressed as a Ramsey sentence, is a sentence in a *simple* observational language. It requires an *extended* observational language, which is observational because it contains no theoretical terms, but has been extended to include an advanced, complicated logic, embracing virtually the whole of mathematics.

Suppose that, in the logical part of this extended observation language, we provide for a series $D_0, D_1, D_2, \ldots$ of domains of mathematical entities such that:

(1) The domain D_0 contains the natural numbers (0, 1, 2, . . .).

(2) For any domain D_n, the domain D_{n+1} contains all classes of elements of D_n.

The extended language contains variables for all these kinds of entities, together with suitable logical rules for using them. It is my opinion that this language is sufficient, not only for formulating all present theories of physics, but also for all future theories, at least for a long time to come. Of course, it is not possible to foresee the kinds of particles, fields, interactions, or other concepts that physicists may introduce in future centuries. However, I believe that such theoretical concepts, regardless of how bizarre and complex they may be, can—by means of Ramsey's device—be formulated in essentially the same extended observation language that is now available, which contains the observational terms combined with advanced logic and mathematics.[2]

On the other hand, Ramsey certainly did not mean—and no one has suggested—that physicists should abandon theoretical terms in their speech and writing. To do so would require enormously complicated statements. For example, it is easy to say in the customary language that a certain object has a mass of five grams. In the symbolic notation of a theory, before it is changed to a Ramsey sentence, one can say that a certain object No. 17 has a mass of five grams by writing, "Mass $(17) = 5$". In Ramsey's language, however, the theoretical term

---

[2] I have defended this view at greater length, and with more technical details, in my paper, "Beobachtungssprache und theoretische Sprache," *Dialectica*, 12 (1958), 236–248; reprinted in W. Ackermann et al., *Logica: Studia Paul Bernays Dedicata* (Neuchâtel (Switzerland): Éditions du Griffon, 1959), pp. 32–44.

"Mass" does not appear. There is only the variable (as in the previous example) "R_3". How can the sentence "Mass (17) = 5" be translated into Ramsey's language? "R_3 (17) = 5" obviously will not do; it is not even a sentence. The formula must be supplemented by the assumptions concerning the relation R_3 that are specified in the Ramsey sentence. Moreover, it would not be sufficient to pick out only those postulate-formulas containing "R_3". *All* the postulates are needed. Therefore, the translation of even this brief sentence into the Ramsey language demands an immensely long sentence, which contains the formulas corresponding to all the theoretical postulates, all the correspondence postulates, and their existential quantifiers. Even when the abbreviated form used earlier is adopted, the translation is rather long:

$$(\exists\, C_1)\ (\exists\, C_2)\ \ldots\ (\exists\, R_3)\ (\exists\, R_4)\ [\ldots\ C_1 \ldots$$
$$C_2 \ldots R_1 \ldots R_2 \ldots R_3 \ldots R_4 \ldots ; \ldots R_1 \ldots$$
$$O_1 \ldots O_2 \ldots O_3 \ldots R_2 \ldots O_4 \ldots O_m \ldots$$
$$\text{and } R_3(17) = 5].$$

It is evident that it would be inconvenient to substitute the Ramsey way of speaking for the ordinary discourse of physics in which theoretical terms are used. Ramsey merely meant to make clear that it was *possible* to formulate any theory in a language that did not require theoretical terms but that said the same thing as the conventional language.

When we say it "says the same thing", we mean this only so far as all observable consequences are concerned. It does not, of course, say *exactly* the same thing. The former language presupposes that theoretical terms, such as "electron" and "mass", point to something that is somehow *more* than what is supplied by the context of the theory itself. Some writers have called this the "surplus meaning" of a term. When this surplus meaning is taken into account, the two languages are certainly not equivalent. The Ramsey sentence represents the full *observational content* of a theory. It was Ramsey's great insight that this observational content is all that is needed for the theory to function as theory, that is, to explain known facts and predict new ones.

It is true that physicists find it vastly more convenient to talk in the shorthand language that includes theoretical terms, such as "proton", "electron", and "neutron". But if they are asked whether electrons "really" exist, they may respond in different ways. Some physicists are content to think about such terms as "electron" in the Ramsey way. They evade the question about existence by stating that there are certain observable events, in bubble chambers and so on, that can be described by certain mathematical functions, within the framework of a certain theoretical system. Beyond that they will assert nothing. To ask whether there really *are* electrons is the same—from the Ramsey point of view—as asking whether quantum physics is true. The answer is that, to the extent that quantum physics has been confirmed by tests, it is justifiable to say that there are instances of certain kinds of events that, in the language of the theory, are called "electrons".

With respect to the nature of theories and the entities referred to in theories, there are at present two main views, often labeled "instrumentalism" and "realism".[3] The instrumentalist point of view is close to the position defended by Charles Peirce, John Dewey, and other pragmatists, as well as by many other philosophers of science. From this point of view, theories are not about "reality". They are simply language tools for organizing the observational phenomena of experience into some sort of pattern that will function efficiently in predicting new observables. The theoretical terms are convenient symbols. The postulates containing them are adopted because they are useful, not because they are "true". They have no surplus meaning beyond the way in which they function in the system. It is meaningless to talk about the "real" electron or the "real" electromagnetic field.

Opposed to this view is the "descriptive" or "realist" view of theories. (Sometimes these two are distinguished, but it is not necessary to delve into these subtle differences.) Advocates of this approach find it both convenient and psychologically comforting to think of electrons, magnetic fields, and gravitational waves as actual entities about which science is steadily learning more. They point out that there is no sharp line separating an observable, such as an apple, from an unobservable, such as a neutron. An amoeba is not observable by the naked eye, but it is observable through a light microscope. A virus is not observable even through a light microscope, but its structure can be seen quite distinctly through an electron microscope. A proton cannot be observed in this direct way, but its track through a bubble chamber can be observed. If it is permissible to say that the amoeba is "real", there is no reason why it is not permissible to say that the proton is

[3] An illuminating discussion of the two or three points of view on this controversy is given by Ernest Nagel, *The Structure of Science* (New York: Harcourt, Brace & World, 1961), Chapter 6, "The Cognitive Status of Theories."

equally real. The changing view about the structure of electrons, genes, and other things does not mean that there is not something "there", behind each observable phenomenon; it merely indicates that more and more is being learned about the structure of those entities.

Proponents of the descriptive view remind us that unobservable entities have a habit of passing over into the observable realm as more powerful instruments of observation are developed. At one time, "virus" was a theoretical term. The same is true of "molecule". Ernst Mach was so opposed to thinking of a molecule as an existing "thing" that he once called it a "valueless image". Today, even atoms in a crystal lattice can be photographed by bombarding them with elementary particles; in a sense, the atom itself has become an observable. Defenders of this view argue that it is as reasonable to say that an atom "exists" as it is to say that a distant star, observable only as a faint spot of light on a long-exposed photographic plate, exists. There is, of course, no comparable way to observe an electron. But that is no reason for refusing to say it exists. Today, little is known about its structure; tomorrow a great deal may be known. It is as correct, say the advocates of the descriptive approach, to speak of an electron as an existing thing as it is to speak of apples and tables and galaxies as existing things.

It is obvious that there is a difference between the meanings of the instrumentalist and the realist ways of speaking. My own view, which I shall not elaborate here, is essentially this. I believe that the question should not be discussed in the form: "Are theoretical entities real?" but rather in the form: "Shall we prefer a language of physics (and of science in general) that contains theoretical terms, or a language without such terms?" From this point of view the question becomes one of preference and practical decision.[4]

---

[4] In my view greater clarity often results if discussions of whether certain entities are real are replaced by discussions of preference of language forms. This view is defended in detail in my "Empiricism, Semantics, and Ontology," *Revue internationale de philosophie,* 4 (1950), 20-40. The paper is reprinted in Philip Wiener, *Readings* (see the Bibliography).

CHAPTER 27

Analyticity in an Observation Language

ONE OF THE OLDEST, most persistent dichotomies in the history of philosophy is that between analytic and factual truth. It has been expressed in many different ways. Kant introduced the distinction, as shown in Chapter 18, in terms of what he called "analytic" and "synthetic" statements. Earlier writers spoke of "necessary" and "contingent" truth.

In my opinion, a sharp analytic-synthetic distinction is of supreme importance for the philosophy of science. The theory of relativity, for example, could not have been developed if Einstein had not realized that the structure of physical space and time cannot be determined without physical tests. He saw clearly the sharp dividing line that must always be kept in mind between pure mathematics, with its many types of logically consistent geometries, and physics, in which only experiment and observation can determine which geometries can be applied most usefully to the physical world. This distinction between analytic truth (which includes logical and mathematical truth) and factual truth is equally important today in quantum theory, as physicists explore the nature of elementary particles and search for a field theory that will bind quantum mechanics to relativity. In this chapter and in the next, we shall be concerned with the question of how this ancient distinction can be made precise throughout the entire language of modern science.

For many years, it has been found useful to divide the terms of a scientific language into three main groups.

1. Logical terms, including all the terms of pure mathematics.
2. Observational terms, or O-terms.
3. Theoretical terms, or T-terms (sometimes called "constructs").

It is true, of course, as has been emphasized in earlier chapters, that no sharp boundary separates the O-terms from the T-terms. The choice of an exact dividing line is somewhat arbitrary. From a practical point of view, however, the distinction is usually evident. Everyone would agree that words for properties, such as "blue", "hard", "cold", and words for relations, such as "warmer", "heavier", "brighter", are O-terms, whereas "electric charge", "proton", "electromagnetic field" are T-terms, referring to entities that cannot be observed in a relatively simple, direct way.

With respect to sentences in the language of science, there is a similar three-fold division.

1. Logical sentences, which contain no descriptive terms.
2. Observational sentences, or O-sentences, which contain O-terms but no T-terms.
3. Theoretical sentences, or T-sentences, which contain T-terms. T-sentences, however, are of two types:
   a. Mixed sentences, containing both O- and T-terms, and
   b. Purely theoretical sentences, containing T-terms but no O-terms.

The entire language, L, of science is conveniently divided into two parts. Each contains the whole of logic (including mathematics). They differ only with respect to their descriptive, nonlogical elements.

1. The observation language, or O-language (L_O), containing logical sentences and O-sentences, but no T-terms.
2. The theoretical language, or T-language (L_T), containing logical sentences and T-sentences (with or without O-terms in addition to T-terms).

The T-terms are introduced into the language of science by a theory, T, which rests upon two kinds of postulates—the theoretical, or T-postulates, and the correspondence, or C-postulates. The T-postulates are the laws of the theory. They are pure T-sentences. The C-postulates, the correspondence rules, are mixed sentences, combining T-terms with O-terms. As shown earlier, they constitute what Campbell called the dictionary for joining the observational and theoretical languages, what Reichenbach called coordinating definitions, and what in Bridgman's terminology might be called operational postulates or operational rules.

With this background, let us turn to the problem of distinguishing between analytic and factual truth in the observational language.

The first kind of analytic truth is logical truth or "L-truth" in our terminology. A sentence is L-true, when it is true in virtue of its form and of the meanings of the logical terms occurring in it. For example, the sentence, "If no bachelor is a happy man, then no happy man is a bachelor", is L-true, because you can recognize its truth if you know the meanings, the way of using the logical words "if", "then", "no", and "is", even if you do not know the meanings of the descriptive words "bachelor", "happy", and "man". All the statements (principles and theorems) of logic and mathematics are of this kind. (That pure mathematics is reducible to logic was shown by Frege and Russell, although some points of this reduction are still controversial. This question will not be discussed here.)

On the other hand, as Willard V. O. Quine has made clear, the observational language is rich in sentences that are analytic in a much wider sense than L-true. These sentences cannot be described as true or false until the meanings of their descriptive terms are understood as well as the meanings of their logical terms. Quine's well-known example is, "No bachelor is married." The truth of this sentence is patently not a matter of the contingent facts of the world, yet it cannot be called true because of its logical form alone. In addition to knowing the meaning of "no" and "is", it is necessary to know what is meant by "bachelor" and "married". In this case, everyone who speaks English would agree that "bachelor" has the same meaning as "a man who is not married". Once these meanings are accepted, it is immediately apparent that the sentence is true, not because of the nature of the world, but because of the meanings our language assigns to the descriptive words. It is not even necessary to understand these meanings fully. It is necessary only to know that the two words have incompatible meanings, that a man cannot be described simultaneously as both a bachelor and a married man.

Quine proposed, and I follow his proposal, that the term "analytic" be used for "logically true" in the broader sense, the sense that includes sentences of the type just discussed, as well as L-true sentences. "A-truth" is the term I use for analytic truth in this broad sense. Thus, all L-true sentences are A-true, although not all A-true sentences are L-true. An L-true sentence is true because of its logical form alone. An A-true sentence, not L-true, is true because of the meanings assigned to its descriptive terms as well as because of the meanings of its logical terms. In contrast, the truth or falsity of a synthetic sentence is not determined by the meanings of its terms, but by factual information about the physical world. "Objects fall to the earth with an acceleration of 32 feet per second per second." It cannot be decided whether the statement is true or false simply by an examination of its meaning. An empirical test is necessary. Such a statement has "factual content". It tells something about the actual world.

Of course, no natural language, such as English, is so precise that everyone understands every word in the same way. For this reason, it is easy to formulate sentences that are ambiguous with respect to their analyticity; they are sentences whose analyticity or syntheticity will be argued about.

Consider, for instance, the assertion, "All red-headed woodpeckers have red heads." Is it analytic or synthetic? At first you may answer that it is, of course, analytic. "Red-headed woodpeckers" *means* "woodpeckers that have red heads", so the sentence is equivalent to the assertion that all woodpeckers with red heads have red heads. Such a sentence is not only A-true but also L-true.

You are right *if* the meaning of "red-headed woodpecker" is such that "having a red head" is, in fact, an essential component of the meaning. But is it an essential component? An ornithologist may have a different understanding of "red-headed woodpecker". For him the term may refer to a species of bird defined by a certain type body structure, shape of bill, and behavior habits. He may consider it quite possible that this species of bird, in some isolated region, may have undergone a mutation that changed the color of its head to, say, white. For sound taxonomic reasons, he would continue to call such birds red-headed woodpeckers even though their heads were not red. They would be a species variant. He might even refer to them as "white-headed red-headed woodpeckers". Therefore, if "red-headed woodpecker" is so interpreted that having a red head is *not* an essential component, the sentence becomes synthetic. It is necessary to make an empirical survey of all red-headed woodpeckers to determine whether all of them do, in fact, have red heads.

Even the statement "If Mr. Smith is a bachelor, he does not have a wife" could be taken as synthetic by anyone who interpreted certain words in an unorthodox way. For example, to a lawyer the word "wife" may have a broad meaning that includes "common-law wife". If a lawyer interprets "bachelor" to mean a man not legally married but takes "wife" in this broader sense, then clearly the sentence is synthetic. One must investigate Mr. Smith's private life to find out whether the sentence is true or false.

The problem of analyticity can be discussed with respect to an artificial observational language that can be constructed by laying down precise rules. These rules need not specify the full meanings of all descriptive words in the language, but meaning relations between certain words must be made clear by rules that I once called "meaning postulates" but now prefer to call, more simply, "A-postulates" (analyticity postulates). We can easily imagine how complete specifications *could* be given for all the language's descriptive words. For example, we could specify the meanings of "animal", "bird", and "red-headed woodpecker" by the following designation rules:

(D1) The term "animal" designates the conjunction of the following properties (1) . . . , (2) . . . , (3) . . . , (4) . . . , (5) . . . (here a complete list of definitory properties would be given).

(D2) The term "bird" designates the conjunction of the following properties (1) . . . , (2) . . . , (3) . . . , (4) . . . , (5) . . . (as in D1 above), plus the additional properties (6) . . . , (7) . . . , (8) . . . , (9) . . . , (10) . . . (all the properties needed to specify the meaning of "bird").

(D3) The term "red-headed woodpecker" designates the conjunction of the following properties (1) . . . , (2) . . . , . . . , (5) . . . (as in D1), plus (6) . . . , (7) . . . , . . . , (10) . . . , (as in D2), plus the additional properties (11) . . . , (12) . . . , (13) . . . , (14) . . . , (15) . . . (all the properties needed to specify the meaning of "red-headed woodpecker").

If all the required properties were written out in the spaces indicated by dots, it is apparent that the rules would be enormously lengthy and cumbersome. Something like this would be necessary if a full specification of the meanings of all descriptive terms in our artificial language were insisted upon. Fortunately, it is not necessary to go to such tire-some lengths. A-postulates can be limited to specifying the *meaning relations* that hold among the language's descriptive terms. For example, for the three terms just discussed, only two A-postulates are needed.

(*A*1) All birds are animals,
(*A*2) All red-headed woodpeckers are birds.

If the three D-rules are given, the two A-postulates can obviously be derived from them. But, since the D-rules are so cumbersome, it is not necessary to formulate them when the purpose is merely to indicate the analytic structure of a language. Only the A-postulates need be given. They are much simpler, and they provide a sufficient basis for making the distinction between analytic and synthetic statements in the language.

Assume that the artificial language is based on the natural language of English but we wish to give A-postulates to make it possible, in all cases, to determine whether a given sentence in the language is analytic. In some cases, the A-postulates can be obtained by consulting an ordinary English dictionary. Consider the sentence, "If a bottle is tossed out of a window, the bottle is defenestrated." Is this analytic or synthetic? The A-postulate, derived from the dictionary definition, says, "*x* is defenestrated if and only if *x* is tossed out of a window." It is apparent at once that the sentence is A-true. It is not necessary to toss a bottle through a window to see whether it does or does not become defenestrated. The truth of the sentence follows from the meaning relations of its descriptive words, as specified by the A-postulate.

An ordinary dictionary may be precise enough to guide us with respect to some sentences, but will be of little help with respect to others. For example, consider those traditionally ambiguous assertions, "All men are rational animals" and "All men are featherless bipeds." The main difficulty here lies in the great ambiguity of what is meant by "men". In our artificial language, there is no difficulty because the list of our A-postulates settles the matter by fiat. If we desire to interpret "men" in such a way that "rationality" and "animality" are essential meaning components of the word, then "All men are rational" and "All men are animals" are listed among the A-postulates. On the basis of these A-postulates, the statement "All men are rational animals" is A-true. On the other hand, if the A-postulates for "men" refer only to the structure of men's physical bodies, then the statement, "All men are rational animals", is synthetic. If analogous A-postulates are not laid down for the terms "featherless" and "biped", this indicates that in our language "featherlessness" and "bipedality" are not considered essential meaning components of "men". The assertion "All men are featherless bipeds" also becomes synthetic. In our language, a one-legged man would still be called a man. A man who grew feathers on his head would still be called a man.

The important point to understand here is that the more precise the list of A-postulates is made, the more precise a distinction can be made between analytic and synthetic sentences in our language. To the extent that the rules are vague, the constructed language will contain sentences that are hazy with respect to their analyticity. Any haziness that remains—and this point is essential—will not be because of lack of clarity in understanding the dichotomy between analytic and synthetic. It will be because of haziness in understanding the meanings of the descriptive words of the language.

Always bear in mind that the A-postulates, although they may seem to do so, do not tell anything about the actual world. Consider, for example, the term "warmer". We may wish to lay down an A-postulate to the effect that the relation designated by this term is asymmetric. "For any x and any y, if x is warmer than y, then y is not warmer than x." If someone says he has discovered two objects A and B, of such a nature that A is warmer than B, and B is warmer than A, we would not respond by saying: "How surprising! What a wonderful discovery!" We would reply: "You and I must have different understandings of the word 'warmer'. To me it means an asymmetric relation; therefore, the situation you have found cannot be described as you have described it." The A-postulate specifying the asymmetric character of the relation "warmer" is concerned solely with the meaning of the word as it is used in our language. It says nothing whatever about the nature of the world.

In recent years, the view that a sharp distinction can be made between analytic and synthetic statements has been strongly attacked by Quine, Morton White, and others.[1] My own views on this matter are

[1] Quine's attack is in his paper, "Two Dogmas of Empiricism," *Philosophical Review*, 60 (1951), 20–43; reprinted in *From a Logical Point of View* (Cambridge: Harvard University Press, 1953); (New York: Harper Torchbooks, 1963). See also his essay, "Carnap and Logical Truth," in Paul Arthur Schilpp, ed., *The Philosophy of Rudolf Carnap* (La Salle, Ill.: Open Court, 1963), pp. 385–406, and my reply, pp. 915–922. For Morton White's animadversions, see his paper, "The Analytic and Synthetic: An Untenable Dualism," in Sidney Hook, ed., *John Dewey* (New York: Dial, 1950), and Part 2 of White's *Toward Reunion in*

given in two papers reprinted in the appendix of the second edition (1956) of my previously cited book *Meaning and Necessity*. The first of these papers, on "Meaning Postulates", replies to Quine by showing in a formal way (as I have indicated informally here) how the distinction can be made precise for a constructed observation language by the simple expedient of adding A-postulates to the rules of the language. My second paper, "Meaning and Synonymy in Natural Languages", indicates how the distinction can be made, not for an artificial language, but for a commonly used language, such as everyday English. Here the distinction must be based on an empirical investigation of speaking habits. This involves new problems, which are discussed in the paper but which will not be considered here.

So far, analyticity has been discussed only in reference to observation languages: the observation language of everyday life, of science, and the constructed observation language of a philosopher of science. It is my conviction that the problem of distinguishing analytic from synthetic assertions in such languages has, in principle, been solved. Moreover, I believe, and I am convinced that almost all working scientists would agree, that, in the observation language of science, the distinction is a useful one. When, however, we seek to apply the dichotomy to the *theoretical* language of science, we meet formidable difficulties. In Chapter 28 some of these difficulties and a possible way of overcoming them are considered.

Philosophy (Cambridge: Harvard University Press, 1956); (New York: Atheneum paperback, 1963). A list of some important articles replying to Quine will be found in Paul Edwards and Arthur Pap, eds., *A Modern Introduction to Philosophy*, 3rd ed. (New York: Free Press, 1973), pp. 744–745.

Analyticity in a
Theoretical Language

BEFORE EXPLAINING how I believe the analytic-synthetic distinction can be made clear with respect to the theoretical language of science, it is important first to understand the huge difficulties involved and how they spring from the fact that T-terms (theoretical terms) cannot be given complete interpretations. In the observation language, this problem does not arise. It is assumed that all meaning relations between the descriptive terms of the observation language have been expressed by suitable A-postulates, as explained in the preceding chapter. With respect to T-terms, however, the situation is altogether different. There is no complete empirical interpretation for such terms as "electron", "mass", and "electromagnetic field". True, a track in a bubble chamber can be observed and explained as produced by an electron passing through the chamber. But such observations provide only partial, indirect empirical interpretations of the T-terms to which they are linked.

Consider, for example, the theoretical term "temperature", as used in the kinetic theory of molecules. There are C-postulates (correspondence rules) that connect this term with the construction and use of a thermometer, for instance. After a thermometer is put into a liquid, a scale reading is observed. C-postulates join this procedure to the T-term "temperature" in such a way that the scale readings provide a partial interpretation of the term. It is partial because this particular interpretation of "temperature" cannot be used for all the sentences in the theory in which the term occurs. An ordinary thermometer works only within a narrow interval on the temperature scale. There are temperatures below which any test liquid would freeze solid and temperatures above which any test liquid would vaporize. For these temperatures, entirely different methods of measurement must be used. Each method is linked by C-postulates with the theoretical concept of "temperature", but it cannot be said that this exhausts the empirical meaning of "temperature". New observations in the future may yield new C-postulates that will add still further to the empirical interpretation of the concept.

Hempel, in Section 7 of his monograph, "Methods of Concept Formation in Science" (*Encyclopedia of Unified Science,* 1953), has drawn a memorable picture of the structure of a theory.

> A scientific theory might therefore be likened to a complex spatial network: Its terms are represented by the knots, while the threads connecting the latter correspond, in part, to the definitions and, in part, to the fundamental and derivative hypotheses included in the theory. The whole system floats, as it were, above the plane of observation and is anchored to it by rules of interpretation. These might be viewed as strings which are not part of the network but link certain parts of the latter with specific places in the plane of observation. By virtue of those interpretive connections, the network can function as a scientific theory: From certain observational data, we may ascend, via an interpretive string, to some point in the theoretical network, thence proceed, via definitions and hypotheses, to other points, from which another interpretive string permits a descent to the plane of observation.[1]

The problem is to find a way to distinguish, in the language that speaks about this complex network, the sentences that are analytic and those that are synthetic. It is easy to identify L-true sentences, that is, sentences that are true in virtue of their logical form. "If all electrons have magnetic moments and particle x has no magnetic moment, then particle x is not an electron." This sentence is clearly L-true. It is not necessary to know anything about the meanings of its descriptive words

[1] The quotation is from Carl G. Hempel, *International Encyclopedia of Unified Science,* Vol. 2, No. 7: *Fundamentals of Concept Formation in Empirical Science* (Chicago: University of Chicago Press, 1952), pp. 23–38.

to see that it is true. But how is the distinction to be made between sentences that are analytic (true in virtue of the meanings of their terms, including their descriptive terms) and sentences that are synthetic (the truth of which cannot be decided without observing the actual world)?

To recognize analytic statements in a theoretical language, it is necessary to have A-postulates that specify the meaning relations holding among the theoretical terms. A statement is analytic if it is a logical consequence of the A-postulates. It must be true in a way that does not demand observation of the actual world; it must be devoid of factual content. It must be true solely by virtue of the meanings of its terms, just as the observational statement "No bachelor is married" is true by virtue of the meanings of "bachelor" and "married". These meanings can be made precise by rules of the observation language. How can comparable A-postulates be formulated to identify analytic statements in a theoretical language containing theoretical terms for which there are no complete interpretations?

The first thought, perhaps, is that the T-postulates alone might serve as A-postulates. It is true that a deductive theory can be built up by combining T-postulates with logic and mathematics, but the result is an abstract deductive system in which the theoretical terms have not even a partial interpretation. Euclidean geometry is a familiar example. It is an uninterpreted structure of pure mathematics. To become a scientific theory, its descriptive terms must be interpreted, at least partially. This means that its terms must be given empirical meanings, which is done, of course, by correspondence rules that connect its primitive terms with aspects of the physical world. Euclidean geometry is thereby transformed into physical geometry. We say that light moves in "straight lines", cross-hairs in a telescope intersect at a "point", and planets move in "ellipses" around the sun. Until the abstract mathematical structure has been interpreted (at least partially) by C-postulates, the semantic problem of distinguishing analytic from synthetic sentences does not even arise. The T-postulates of a theory cannot be used as A-postulates because they fail to provide the T-terms with empirical meaning.

Can the C-postulates be used to provide A-postulates? The C-postulates cannot, of course, be taken alone. To obtain the fullest possible interpretation (though still only partial) for the T-terms, it is necessary to take the entire theory, with its combined T- and C-postulates. Suppose, then, that we presuppose the entire theory. Will the combined T- and C-postulates furnish the A-postulates we seek? No; now we have presupposed *too much*. We have, indeed, obtained all the empirical meaning we can get for our theoretical terms, but we have also obtained factual information. The conjunction of T- and C-postulates, therefore, gives us synthetic statements, and, as we have seen, such statements cannot provide A-postulates.

An example will make this clear. Suppose we say that the T- and C-postulates of the general theory of relativity will serve as A-postulates for identifying analytic sentences in the theory. On the basis of certain T- and C-postulates, aided by logic and mathematics, we deduce that light from stars will be deflected by the sun's gravitational field. Can we not say that this conclusion is analytic, true solely by virtue of the empirical meanings that have been assigned to all the descriptive terms? We cannot, because the general theory of relativity provides conditional predictions about the world that can be confirmed or refuted by empirical tests.

Consider, for instance, the statement, "These two photographic plates were made of the same pattern of stars. The first was made during a total eclipse of the sun, when the sun's covered disk was inside the star pattern. The second was made when the sun did not appear near this pattern." This will be called statement A. Statement B is: "On the first plate, the images of stars very close to the rim of the eclipsed sun will be displaced slightly from their positions as shown on the second plate and will be displaced in a direction away from the sun." The conditional assertion, "If A, then B", is a statement that can be derived from the general theory of relativity. But it is also a statement that can be tested by observation. Indeed, as shown in Chapter 16, a historic test of this assertion was made by Finlay Freundlich in 1919. He knew that A was true. After careful measurements of the spots of light on the two plates, he found B to be true also. Had he found B false, the conditional statement, "If A, then B", would have been falsified. This, in turn, would have refuted the theory of relativity, from which "If A, then B" was derived. There is, therefore, factual content in the theory's assertion that starlight is deflected by gravitational fields.

To make the same point more formally, after the T- and C-postulates of relativity theory have been specified, it is possible, on the basis of a given set of premises, A, in the observation language, to derive another set of sentences, B, also in the observation language, that cannot be derived without TC, the entire theory. The statement, "If A, then B", is, therefore, a logical consequence of the conjunction of T and C.

If T and C were taken as A-postulates, it would be necessary to look upon the statement "If A, then B" as analytic. But, it clearly is not analytic. It is a synthetic statement in the observation language. It would be falsified if observation of the actual world showed A to be true and B to be false.

Quine and other philosophers of science have argued that the difficulties here are so great that the analytic-synthetic dichotomy, in its intended meaning, cannot be applied to the theoretical language of science. More recently, this view has been presented with great clarity by Hempel.[2] Hempel was willing, perhaps hesitatingly, to accept the dichotomy with respect to the observation language. Regarding its usefulness with respect to the theoretical language, he voiced strong Quinian scepticism. The double role of the T- and C-postulates, he maintained, makes the concept of analytic truth, with respect to a theoretical language, entirely elusive. It is hardly to be imagined, he thought, that a way exists to split up these two functions of the T- and C-postulates so that it can be said that this part of them contributes to meaning, and, therefore, the sentences that rely on that part are, if they are true, true because of meaning only, whereas the other sentences are factual sentences.

One extreme way to solve, or rather avoid, all the troublesome problems connected with theoretical terms is the one proposed by Ramsey. As shown in Chapter 26, it is possible to state the entire observational content of a theory in a sentence known as the Ramsey sentence, ^{R}TC, in which only observational and logical terms appear. The theoretical terms are, it may be said, "quantified away". Since there are no theoretical terms, there is no theoretical language. The problem of defining analyticity for a theoretical language disappears. This, however, is too radical a step. As shown earlier, giving up the theoretical terms of science leads to great complexities and inconveniences. Theoretical terms enormously simplify the task of formulating laws and, for that reason alone, cannot be eliminated from the language of science.

I believe that there is a way to solve the problem by making use of the Ramsey sentence, but only by doing so in a manner that does not force us to take Ramsey's final, extreme step. By making certain distinc-

---

[2] See Hempel's two papers, "The Theoretician's Dilemma" in Herbert Feigl, Michael Scriven, and Grover Maxwell, eds., *Minnesota Studies in the Philosophy of Science* (Minneapolis, Minn.: University of Minnesota Press, 1956), Vol. II, and "Implications of Carnap's Work for the Philosophy of Science" in Paul Arthur Schilpp, ed., *The Philosophy of Rudolf Carnap* (La Salle, Ill.: Open Court, 1963).

tions, the desired dichotomy between analytic and synthetic truth in the theoretical language can be obtained, and, at the same time, all the theoretical terms and sentences of a theory can be retained.

So far, we have regarded a theory as consisting of two "sentences": sentence T, the conjunction of all the T-postulates, and sentence C, the conjunction of all the C-postulates. The theory TC is the conjunction of these two sentences.

I shall propose another way in which the theory TC can be split into two sentences that, taken in conjunction, are equivalent to the theory. It will be divided into a sentence A_T and a sentence F_T. The sentence A_T is intended to serve as the A-postulate for all the theoretical terms of the theory. It must, of course, be completely devoid of factual content. The sentence F_T is intended to be the sentence that expresses the entire observational or factual content of the theory. As has been shown, the Ramsey sentence itself, ^RTC, does exactly this. It expresses, in an observation language extended to include all of mathematics, everything that the theory says about the actual world. It provides no interpretation of theoretical terms, because no such terms appear in the sentence. Thus, the Ramsey sentence, ^RTC, is taken as the factual postulate F_T.

The two sentences F_T and A_T, taken together, should logically imply the entire theory TC. How can a sentence A_T that fulfills these requirements be formulated? For any two sentences S_1 and S_2, the weakest sentence that, together with S_1, logically implies S_2, is the conditional assertion, "If S_1, then S_2". In symbolic form, this is expressed with the familiar symbol for material implication: "$S_1 \supset S_2$". Thus, the simplest way to formulate an A-postulate A_T, for a theory TC, is:

(A_T) $\qquad\qquad\qquad$ $^RTC \supset TC$

It can easily be shown that this sentence is factually empty. It tells nothing about the world. All the factual content is in the sentence F_T, which is the Ramsey sentence ^RTC. The sentence A_T simply asserts that *if* the Ramsey sentence is true, we must then understand the theoretical terms in such a way that the entire theory is true. It is a purely analytic sentence, because its semantic truth is based on the meanings intended for the theoretical terms. This assertion, coupled with the Ramsey sentence itself, will then L-imply the entire theory.

Let us see how this curious A-postulate $^RTC \supset TC$ provides a way to distinguish between analytic and synthetic statements in the theoretical language. The Ramsey sentence ^{R}TC is synthetic. Its truth can be established only by actual observation of the world. But any statement L-implied by the given A-postulate will be analytic.

Here, as with analytic sentences in the observation language, there is a loose sense in which the A-postulate *does* say something about the world. But, in a strict sense, it does not. The A-postulate states that *if* entities exist (referred to by the existential quantifiers of the Ramsey sentence) that are of a kind bound together by all the relations expressed in the theoretical postulates of the theory and that are related to observational entities by all the relations specified by the correspondence postulates of the theory, then the theory itself is true. The A-postulate *seems* to tell something about the world, but it actually does not. It does not tell us whether the theory is true. It does not say that this is the way the world is. It says only that *if* the world is this way, then the theoretical terms must be understood as satisfying the theory.

In Chapter 26, an example of a theory with six theoretical concepts, namely, two classes and four relations, was considered. A schematic formulation (with the context simply indicated by dots) of the theory TC and its Ramsey sentence, ^{R}TC, was given. With this example in view, the A-postulate for this theory can be formulated as follows:

$$(A_T) \quad (\exists C_1)\,(\exists C_2)\,(\exists R_1)\,(\exists R_2)\,(\exists R_3)\,(\exists R_4)\,[\ldots C_1 \ldots$$
$$C_2 \ldots R_1 \ldots R_2 \ldots R_3 \ldots R_4 \ldots ; \ldots R_1$$
$$\ldots O_1 \ldots O_2 \ldots O_3 \ldots R_2 \ldots O_4 \ldots O_m \ldots]$$
$$\supset [\ldots \text{Mol} \ldots \text{Hymol} \ldots \text{Temp} \ldots$$
$$\text{Press} \ldots \text{Mass} \ldots \text{Vel} \ldots ; \ldots \text{Temp} \ldots$$
$$O_1 \ldots O_2 \ldots O_3 \ldots \text{Press} \ldots O_4 \ldots O_m \ldots].$$

This says that, if the world is such that there is at least one sextuple of entities (two classes and four relations) that are related among themselves and to the observational entities, $O_1, O_2, \ldots, O_m$, as specified by the theory, then the theoretical entities Mol, Hymol, Temp, Press, Mass, and Vel form a sextuple that satisfies the theory. It is important to understand that this is not a factual statement asserting that, under the conditions stated, six specified entities do, as a matter of fact, satisfy the theory. The six theoretical terms do not name six specified entities. Before the A-postulates A_T is laid down, these terms have no interpretation, not even a partial one. The only interpretation they receive in this form of the theory is the partial interpretation they obtain *through this A-postulate*. Thus, the postulate says in effect that, if there are one or more sextuples of entities that satisfy the theory, then the six theoretical terms are to be interpreted as denoting six entities forming a sextuple of that kind. If there are, as a matter of fact, sextuples of that kind, then the postulate gives a partial interpretation of the theoretical terms by limiting the sextuples admitted for denotation to the sextuples of that kind. If, on the other hand, there are no sextuples of that kind—in other words, if the Ramsey sentence happens to be false—then the postulate is true irrespective of its interpretation (because, if *"A"* is false, *"A ⊃ B"* is true). Hence, it does not give even a partial interpretation of the theoretical terms.

Once all this is fully grasped, there is no barrier to taking the conditional statement $^{R}TC \supset TC$ as an A-postulate for TC in the same way that A-postulates are taken in the observation language. As an A-postulate in the observation language tells something about the meaning of the term "warmer", so the A-postulate for the theoretical language gives some information about the meaning of theoretical terms, such as "electron" and "electromagnetic field". This information, in turn, permits us to find out that certain theoretical sentences are analytic, namely those that follow from the A-postulate A_T.

It is now possible to say precisely what is meant by A-truth in the total language of science. A sentence is A-true if it is L-implied by the combined A-postulates, that is, by the A-postulates of the observation language together with the A-postulate of any given theoretical language. A sentence is A-false if its negation is A-true. If it is neither A-true nor A-false, it is synthetic.

I use the term "P-truth"—truth based on the postulates—for the kind of truth possessed by sentences if and only if they are L-implied by the postulates, namely, the F-postulate (Ramsey sentence), together with both observational and theoretical A-postulates. In other words, P-truth is based on the three postulates F_T, A_O, and A_T. But, since F_T and A_T together are equivalent to TC, the original form of the theory, it may be just as well to represent all the postulates together as TC and A_O.

On the basis of the various kinds of truth that have been defined and the corresponding kinds of falsity, a general classification of the sentences of a scientific language is obtained. It may be diagrammed as shown in Figure 28–1. This classification cuts across the previous division of the language into logical, observational, theoretical, and mixed sentences given earlier, which are based on the kinds of terms occurring in the sentences. As the reader will note, the traditional term "synthetic" is listed as an alternative to "A-indeterminate"; this seems natural, since the term "A-true" was used for that concept defined as an explication of the customary term "analytic" (or "analytically true"). On the other hand, the term "P-indeterminate" applies to a narrower class, namely, to those A-indeterminate (or synthetic) sentences for which truth or falsity is not even determined by the postulates of the theory *TC*, as, for example, the basic laws of physics or some other field of science. Here the term "contingent" suggests itself as an alternative.

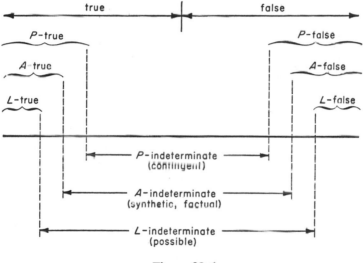

Figure 28–1.

I do not wish to be dogmatic about this program of a classification and, in particular, about the definition of A-truth based upon the proposed A-postulate. Rather, I offer it as a tentative solution to the problem of defining analyticity for the theoretical language. Earlier, although I did not share the pessimism of Quine and Hempel, I always admitted that it was a serious problem and that I could not see a satisfactory solution. For a while I thought we would perhaps have to resign ourselves to taking a sentence that contained theoretical terms and no observation terms as analytic only under the most narrow and almost trivial condition that it is L-true. For example: "Either a particle is an electron, or it is not an electron." Finally, after many years of searching, I found this new approach, with the new A-postulate.[3] No difficulties have yet been discovered in this approach. I am now confident that there is a solution and that, if difficulties appear, it will be possible to overcome them.

[3] A more formal presentation of this approach can be found in my 1958 paper cited in Chapter 26, note 2, and in my reply to Hempel in Paul Arthur Schilpp, ed., *The Philosophy of Rudolph Carnap* (La Salle, Ill.: Open Court, 1963), pp. 958–966.

Part **VI**

BEYOND
DETERMINISM

CHAPTER 29

Statistical Laws

PHILOSOPHERS OF science, in the past, have been very much concerned with the question: "What is the nature of causality?" I have tried to make clear in previous chapters why this is not the best way to formulate the problem. Whatever sort of causality there is in the world is expressed by the laws of science. If we wish to study causality, we can do so only by examining those laws, by studying the ways in which they are expressed and how they are confirmed or disconfirmed by experiment.

In examining the laws of science, it was found to be convenient to distinguish empirical laws, which deal with observables, from theoretical laws, which concern nonobservables. We saw that, although there is no sharp line separating observables from nonobservables, and therefore, no sharp line separating empirical laws from theoretical ones, the distinction is nevertheless useful. Another important and useful distinction, which cuts across both empirical and theoretical laws, is the distinction between deterministic and statistical laws. This distinction has been met before, but, in this chapter, it will be discussed in more detail.

A deterministic law is one that says that, under certain conditions, certain things will be the case. As has been shown, a law of this type may be stated in either qualitative or quantitative terms. The assertion that, when an iron rod is heated, its length increases, is a qualitative assertion. The assertion that, when the bar is heated to a certain temperature, its length increases by a certain amount, is a quantitative assertion. A quantitative deterministic law always states that, if certain magnitudes have certain values, another magnitude (or one of the former magnitudes at a different time) will have a certain value. In brief, the law expresses a functional relation between the values of two or more magnitudes.

A statistical law, however, states only a probability distribution for the values of a magnitude in individual cases. It gives only the average value of a magnitude in a class of many instances. For example, a statistical law states that, if a cubical die is rolled sixty times, a given face may be expected to be uppermost on about ten of the rolls. The law does not predict what will happen on any one roll, nor does it say what is certain to happen on sixty rolls. It asserts that, if a great many rolls are made, each face can be expected to appear about as often as any other face. Because there are six equally probable faces, the probability of rolling any one face is $\frac{1}{6}$. Probability is used here in a statistical sense, to mean relative frequency in the long run, and not in the logical or inductive sense, which I call degree of confirmation.

Statistical laws were common enough in the nineteenth century, but no physicist then imagined that such laws indicated an absence of determinism in the basic laws of nature. He assumed that statistical laws were made either for reasons of convenience or because sufficient knowledge was not available to describe a situation in a deterministic way.

Statements issued by a government, after a population census, are familiar examples of statements expressed in statistical form for reasons of convenience rather than of ignorance. During a census, the government attempts to obtain from every individual a record of his age, sex, race, place of birth, number of dependents, state of health, and so forth. By carefully counting all these facts, the government is able to issue valuable statistical information. (In previous times, the counting and calculation were done by hand. There was usually a ten-year interval from one census to the next, and, by the time a new census started, the calculations had not been completed on the old one. Today the data are put on punch cards, and computers do the work rapidly.) The data reveal that a certain percentage of individuals are over sixty years of age, a certain percentage are doctors, a certain percentage have tuberculosis, and so forth. Statistical statements of this sort are necessary in order to reduce an enormous number of facts to manageable form. This does not mean that the individual facts are not available; it means only that it would be extremely inconvenient to express them as individual facts. Instead of making millions of single statements, such as, ". . . and there is also Mrs. Smith, of San Francisco, who was born in Seattle, Washington, is seventy-five years old, has four children and ten grand-children", the information is compressed into short statistical statements. This is done for reasons of convenience, even though all the underlying facts are on record.

Sometimes the single facts are not available, but it is possible to obtain them. For example, instead of taking a complete census of every individual in a large population, only a representative sample may be investigated. If the sample indicates that a certain percentage of the population own their own homes, it may be concluded that approximately the same percentage of the entire population own homes. It would be possible to check every individual, but, rather than go to the time and expense of such an undertaking, a sample check is made. If the sample is chosen carefully, so that there is good reason to regard it as representative, it is possible to obtain good general estimates.

Even in the physical and biological sciences, it is frequently convenient to make statistical statements, although individual facts are known or would not be difficult to obtain. A plant breeder may disclose that approximately a thousand plants with red blossoms were subjected to certain conditions; in the next generation of plants, about 75 per cent of the blossoms were white instead of red. The botanist may know the exact numbers of red and white blossoms, or, if he does not, it may be possible for him to obtain the numbers by making exact counts. But, if there is no need for such accuracy, he may find it more convenient to express the results as a rough percentage.

At times it is extremely difficult, even impossible, to obtain exact information about individual cases, although it is easy to see how it *could* be obtained. For instance, if we could measure all the relevant magnitudes involved in the fall of a die—its exact position at the time it leaves the hand, the exact velocities imparted to it, its weight and elasticity, the nature of the surface on which it bounces, and so on—it would be possible to predict exactly how the die would come to rest. Because machines for making such measurements are not presently available, we must be content with a statistical law expressing a long run frequency.

In the nineteenth century, the kinetic theory of gases led to the formulation of many probabilistic laws in the field known as statistical mechanics. If a certain quantity of, say, oxygen has a certain pressure and temperature throughout, there will be a certain distribution of the velocity of its molecules. This is called the Maxwell-Boltzmann distribution law. It says that, for each of the three components of the velocity, the probability distribution is the so-called normal (or Gaussian) function, represented by the familiar bell-shaped curve. It is a statistical law about a situation in which the facts concerning each individual molecule were technically impossible to obtain. Here the ignorance—and this point is important—is deeper than the ignorance involved in previous examples. Even in the case of the die, it is conceivable that instruments could be built for analyzing all the relevant facts. The facts could be fed to an electronic computer, and, before the die stopped rolling, the computer would flash: "It will be a six-spot." But, with regard to the molecules of a gas, there is no known technique by which the direction and velocity of each individual molecule can be measured and the billions of results then analyzed in order to check whether the Maxwell-Boltzmann distribution law holds. Physicists formulated that law as a microlaw, expressed in the theory of gases and confirmed by testing various consequences derived from the law. Such statistical laws were common, in the nineteenth century, in fields in which it was impossible to obtain individual facts. Today, laws of this type are used in every branch of science, especially in the biological and social sciences.

Nineteenth-century physicists were fully aware that the probabilistic laws of gases or laws respecting human behavior concealed an ignorance more profound than the ignorance involved in the throw of a die. Nevertheless, they were convinced that *in principle* such information was not impossible to obtain. To be sure, no technical means were on hand for measuring individual molecules, but that was just an unfortunate limitation to the power of available tools. Under a microscope, the physicist could see small particles, suspended in a liquid and dancing about erratically as they were pushed this way and that by collisions with invisible molecules. With better instruments, smaller and smaller particles could be observed. In the future, perhaps, instruments might be built to measure the positions and velocities of individual molecules.

There are, of course, serious optical limitations. The nineteenth-century physicists also knew that, when a particle is no larger than the wave length of visible light, it is not possible to see it in any conceivable type of light microscope. But this did not rule out the possibility of other types of instruments that might measure particles smaller than the wave length of light. Indeed, today's electron microscopes enable one to "see" objects that are below the theoretical limit of optical microscopes. Nineteenth-century scientists were convinced that there is in principle no limit to the precision with which observations of smaller and smaller objects can be made.

They realized, also, that no observation is ever *completely* precise. There is always an element of uncertainty. *All* laws of science are, in this sense, statistical; but it is a trivial sense. The important point is that precision can always be increased. Today, said the nineteenth-century physicists, it is possible to measure something with a precision of two decimal digits. Tomorrow it will be possible to reach a precision of three decimal digits, and, decades from now, perhaps we will reach twenty- or a hundred-digit precision. There seemed to be no limit to the precisions that could be obtained in any type of measurement. Nineteenth-century physicists and many philosophers as well took for granted that, behind all the macrolaws, with their inescapable errors of measurement, are microlaws that are exact and deterministic. Of course, actual molecules cannot be seen. But surely, if two molecules collide, their resulting motions will be completely determined by conditions prior to the collision. If all these conditions could be known, it would be possible to predict exactly how the colliding molecules would behave. How could it be otherwise? The behavior of molecules must depend on *something*. It cannot be arbitrary and haphazard. The basic laws of physics must be deterministic.

Nineteenth-century physicists also recognized that basic laws are idealizations seldom exemplified in pure form because of the influence of extraneous factors. They expressed this by distinguishing between basic laws and "restricted" laws, which derive from the basic laws. A restricted law is simply a law formulated with a restricting clause; it says, for example, that this or that will happen only under "normal circumstances". Consider: "An iron rod heated from freezing temperature to that of boiling water will increase in length." This is not true if the rod is clamped in a strong vise that exerts pressure on the ends. If the pressure is sufficient, expansion of the rod is prevented. The law is restricted, therefore, in the sense that it is understood to hold only under normal circumstances, that is, when no other forces act on the rod to disturb the experiment.

Behind all the restricted laws are the fundamental laws that make unconditional assertions. "Two bodies attract each other with a gravitational force proportional to each of their masses and inversely proportional to the square of the distance between them." This is an unconditional statement. There could, of course, be other forces, such as magnetic attraction, that could change the motion of one of the two bodies but that would not change the amount or direction of the gravitational force. No restrictive clauses need be added to the statement of the law. Another example is provided by Maxwell's equations for the electromagnetic field. They were understood to hold unconditionally, with absolute precision. The great picture presented by Newtonian physics was that of a world in which all events could, in principle, be explained by basic laws that were completely free of indeterminacy. As shown in a previous chapter, Laplace gave a classic formulation of this view by saying that an imaginary mind, which knew all the fundamental laws and all the facts about the world at one instant in its history, would be able to calculate all the world's past and future events.

This utopian picture was shattered by the rise of quantum physics, as we shall see in the next and final chapter.

CHAPTER 30

Indeterminism in
Quantum Physics

THE ESSENTIALLY nondeterministic character of quantum mechanics rests on the principle of indeterminacy, sometimes called the uncertainty principle, or uncertainty relation, first stated in 1927 by Werner Heisenberg. It says, roughly, that, for certain pairs of magnitudes called "conjugate" magnitudes, it is impossible in principle to measure both at the same instant with high precision.

An example of such a pair is:

(1) The x-coordinate (q_x) of the position of a given particle at a given time (with respect to a given coordinate system).

(2) The x-component (p_x) of the momentum of the same particle at the same time. (This component is the product of the mass of the particle and the x-component of its velocity.)

The same holds for the pair q_y, p_y, and for the pair q_z, p_z.

Suppose that measurements are made of two conjugate magnitudes p and q and that it is found that p lies within a certain interval of length $\triangle p$ and that q lies within a certain interval of length $\triangle q$. Heisenberg's uncertainty principle asserts that, if we try to measure p precisely, that is, make $\triangle p$ very small, we cannot at the same instant measure q precisely, that is, make $\triangle q$ very small. More specifically, the product of $\triangle p$ and $\triangle q$ cannot be made smaller than a certain value which is expressed in terms of Planck's quantum constant h. If the conjugate magnitudes are components of momentum and position, the uncertainty principle says that it is not possible in principle to measure both with a high degree of accuracy. If we know exactly where a particle is, its momentum components become hazy. And if we know exactly what its momentum is, we cannot pin down exactly where it is. In actual practice, of course, the inaccuracy of a measurement of this kind is usually much larger than the minimum given by the uncertainty principle. The important point, the implications of which are enormous, is that this inaccuracy is part of the basic laws of quantum theory. The limitation stated by the uncertainty principle must not be understood as being due to the imperfections of measuring instruments and therefore as something that can be reduced by improvements in measuring techniques. It is a fundamental law that must hold as long as the laws of quantum theory are maintained in their present form.

This does not mean that the laws accepted in physics cannot be changed or that Heisenberg's uncertainty principle will never be abandoned. Nevertheless, I believe it is fair to assert that it would take a revolutionary change in the basic structure of present-day physics to remove this feature. Some physicists today are convinced (as was Einstein) that this feature of modern quantum mechanics is questionable and may some day be discarded. That is a possibility. But the step would be a radical one. At the moment, no one can see how the uncertainty principle can be eliminated.

A related and equally important difference between quantum theory and classical physics lies in the concept of an instantaneous state of a physical system. Consider, as an example, a physical system consisting of a number of particles. In classical physics, the state of this system at the time t_1 is completely described by giving for each particle the values of the following magnitudes (sometimes called "state variables", I shall call them "state magnitudes"):

(a) The three position coordinates at t_1.
(b) The three momentum components at t_1.

Assume that this system remains isolated during the time from t_1 to t_2; that is to say, it is not affected during this time interval by any disturbance from outside. Then, on the basis of the given state of the system at t_1, the laws of classical mechanics determine uniquely its state (the values of all the state magnitudes) at t_2.

The picture in quantum mechanics is entirely different. (We shall disregard here the difference in the nature of those particles which are regarded as ultimate in the sense of being indivisible. In modern physics this character is no longer ascribed to atoms, but to smaller particles, such as electrons and protons. Although this difference marks a great step forward in the recent development of physics, it is not essential for our present discussion of the formal methods of specifying the state of a system.) In quantum mechanics a set of state magnitudes for a given system at a given time is called a "complete" set if, first, it is in principle possible to measure all the magnitudes of the set simultaneously and if, second, for any other state magnitude which may be measured simultaneously with all of those in the set, its value is determined by their values. Thus, in our example of a class of particles, a complete set may consist of the following magnitudes: for some of the particles, the coordinates q_x, q_y, and q_z; for some other particles, the momentum components p_x, p_y, p_z; for still others, p_x, q_y, p_z or q_x, q_y, p_z; and for yet other particles, other suitable sets of three magnitudes expressed in terms of the q's and p's. According to the principles of quantum mechanics, the state of a system at a given time is completely described by specifying the values of any complete set of state magnitudes. Evidently, such a description would be regarded as incomplete from the classical point of view, because, if the set contains q_r, then p_x is neither given nor determined by the other values in the set. But this restriction of a state description is in line with the uncertainty principle: if q_x is known, p_x is in principle unknowable. It is easily seen that there is an enormous number—indeed an infinite number—of different possible choices of a complete set of state magnitudes for a given system. We may freely choose to make measurements of the magnitudes of any *one* of the complete sets. And after having measured the exact values of the magnitudes of the chosen set, then the state description specifying those values is the one we may claim to know.

In quantum mechanics, any state of a system can be represented by a function of a special kind called a "wave function". A function of this kind assigns numerical values to the points of a space. (This is not, however, in general, our familiar three-dimensional space, but an abstract higher-dimensional space. If the values of a complete set of state magnitudes for the time t_1 are given, the wave function of the system for t_1 is uniquely determined. These wave functions, although each is based on a set of magnitudes that would appear incomplete from the point of view of classical physics, play in quantum mechanics a role analogous to that of the state descriptions in classical mechanics. Under the condition of isolation as before, it is possible to determine the wave function for t_2 on the basis of the given wave function for t_1. This is done with the help of a famous equation known as the "Schrödinger differential equation", first stated by Erwin Schrödinger, the great Austrian physicist. This equation has the mathematical form of a deterministic law; it yields the complete wave function for t_2. Therefore, if we accept the wave functions as complete representations of instantaneous states, we would be led to say that, at least on the theoretical level, determinism is preserved in quantum physics.

Such an assertion, although made by some physicists, seems to me misleading because it might induce the reader to overlook the following fact. When we ask what the wave function calculated for the future time point t_2 tells us about the values of the state magnitudes at t_2, the answer is: if we plan to make at t_2 a measurement of a particular state magnitude—for example, the y-coordinate of the position of particle number 5—then the wave function does not predict the value that our measurement will find; it supplies only a probability distribution for the possible values of this magnitude. In general, the wave function will assign positive probabilities to several possible values (or to several sub-intervals of possible values). Only in some special cases does one of the values theoretically reach a probability of 1 (certainty), permitting us to say that the value has been definitely predicted. Note that the wave function calculated for t_2 supplies a probability distribution for the values of *every* state magnitude of the physical system under consideration. In our earlier example, this means that it supplies probability distributions for all the magnitudes mentioned under both (a) and (b). Quantum theory is fundamentally indeterministic in that it does not supply definite predictions for the results of measurements. It supplies only probability predictions.

Because the wave function calculated for time t_2 yields probability distributions for the primary state magnitudes with respect to single particles, it is likewise possible to derive probability distributions for other magnitudes that are defined in terms of the primary ones. Among these other magnitudes are the statistical magnitudes with respect to the set of all the particles of the physical system, or a subset of these particles. Many of these statistical magnitudes correspond to macro-observable properties; for example, to the temperature of a small, but visible body or to the position or velocity of a body's center of gravity. If the body is made up of billions of particles—for example, an artificial satellite circling the earth—its position, velocity, temperature, and other measurable magnitudes can be calculated with great accuracy. In cases of this kind, the probability density curve for a statistical magnitude has the shape of an extremely narrow, steep hill. We can specify, therefore, a small interval which includes practically the entire hill; as a consequence, the probability of the event that the value of the magnitude lies in this interval is extremely close to 1. It is so close that, for all practical purposes, we may disregard the probability character of the prediction and take it as if it were certain. But from the standpoint of quantum theory, the satellite is a system made up of billions of particles, and, for each individual particle there is an inescapable fuzziness in the predictions. The uncertainty expressed by the quantum laws holds also for the satellite, but is reduced almost to zero by the statistical laws covering the very large number of particles.

On the other hand, there are situations of a quite different nature in which the occurrence of an event is directly observable in the strongest sense, but is nevertheless dependent on the behavior of an extremely small number of particles; sometimes, even a single particle. In cases of this kind, the considerable uncertainty with respect to the particle's behavior holds likewise for the macroevent. This occurs often in those situations where a radioactive microevent 'triggers" a macroevent; for example, when an electron emitted in beta-decay produces a clearly audible click in a Geiger counter. Even if we make the idealized assumption that we know the values of a complete set of primary state magnitudes for the subatomic particles in a small set of radioactive atoms constituting the body B at time t_1, we could derive only probabilities for the occurrence of such events as: no emitted particle, one emitted particle, two emitted particles, and so on, within the first second following t_1. If the process is such that the probability of no emission in the one-second interval is near to 1, we cannot predict, even with crude approximation, the time at which the first particle emission will take place and cause the Geiger counter click. We can only determine probabilities and related values; for example, the expectation value of the time of the first click.

In view of this situation, I would say that nineteenth-century determinism has been discarded in modern physics. I believe that most physicists today would prefer this way of expressing the radical alteration that quantum mechanics has made in the classical Newtonian picture.

When some philosophers, such as Ernest Nagel, and some physicists, such as Henry Margenau, say that there is still determinism in the laws about the states of systems and that only the definition of "state of a system" has changed, I would not oppose their view. What they say is indeed the case. But, in my opinion, the word "only" can be misleading. It gives the impression that the change is merely a different answer to the question: What are the magnitudes that characterize the state of a system? Actually, the change is much more fundamental. Classical physicists were convinced that, with the progress of research, laws would become more and more exact, and that there is no limit to the precision that can be obtained in predicting observable events. In contrast, the quantum theory sets an insuperable limit. For this reason, I think there is less risk of misunderstanding if we say that the causality structure—the structure of laws—in modern physics is fundamentally different from what it was from the time of Newton to the end of the nineteenth century. Determinism in the classical sense has been abandoned.

It is easy to understand why this radically new picture of physical law was at first psychologically difficult for physicists to accept.[1] Planck himself, by nature a conservative thinker, was distressed when he first realized that the emission and absorption of radiation was not a continuous process, but rather one that proceeded in indivisible units. This discreteness was so entirely against the whole spirit of traditional physics that it was extremely difficult for many physicists, including Planck, to adjust to the new way of thinking.

The revolutionary nature of the Heisenberg uncertainty principle has led some philosophers and physicists to suggest that certain basic changes be made in the language of physics. Physicists themselves seldom talk much about the language they use. Such talk usually comes only from those few physicists who are also interested in the logical

[1] On this point I would recommend a little book by Werner Heisenberg called *Physics and Philosophy: The Revolution in Modern Science* (New York: Harper, 1958). It contains a clear account of the historical development of quantum theory—the first hesitant steps by Planck, then the contributions of Einstein, Heisenberg, and others. F. S. C. Northrop correctly points out, in his introduction, that Heisenberg is much too modest in discussing his own role in this history.

foundations of physics or from logicians who have studied physics. Those people ask themselves: "Should the language of physics be modified to accommodate the uncertainty relations? If so, how?"

The most extreme proposals for such modification concern a change in the form of logic used in physics. Philipp Frank and Moritz Schlick (Schlick was then a philosopher in Vienna and Frank was a physicist in Prague) together first expressed the view that, under certain conditions, the conjunction of two meaningful statements in physics should be considered meaningless; for example, two predictions concerning the values of conjugate magnitudes for the same system at the same time. Let statement *A* predict the exact position coordinates of a particle for a certain time point. Let statement *B* give the three momentum components of that same particle for the same time point. We know, from the Heisenberg uncertainty principle, that we have only two choices:

1. We can make an experiment by which we learn (provided, of course, we have sufficiently good instruments) the position of a particle with high, though not perfect, precision. In this case, our determination of the particle's momentum will be highly unprecise.

2. We can instead make another experiment by which we measure the momentum components of the particle with great precision. In this case, we must be content with a highly imprecise determination of the particle's position.

In short, we can test either for *A* or for *B*. We cannot test for the conjunction "*A* and *B*". Martin Strauss, a pupil of Frank, wrote his doctoral dissertation on this and related problems. Later, he worked with Niels Bohr, in Copenhagen. Strauss held that the conjunction of *A* and *B* should be taken as meaningless, because it is not confirmable. We can verify *A*, if we wish, with any desired precision. We can do the same for *B*. We cannot do it for "*A* and *B*". The conjunction should not, therefore, be considered a meaningful statement. For this reason, Strauss maintained, the formation rules (rules specifying the admitted forms of sentences) of the language of physics should be modified. In my view, such a radical change is inadvisable.

Another, similar suggestion was advanced by the mathematicians Garrett Birkhoff and John von Neumann.[2] They suggested a change, not in the formation rules, but in the transformation rules (rules by which a sentence may be derived from another sentence or set of sen-

[2] See Garrett Birkhoff and John von Neumann, "The Logic of Quantum Mechanics," *Annals of Mathematics*, 37 (1936), 823–843.

tences). They proposed that physicists abandon one of the laws of distribution in propositional logic.

A third proposal was made by Hans Reichenbach, who suggested that the traditional two-valued logic be replaced by a three-valued logic.[3] In such a logic, each statement would have one of three possible values: T (true), F (false), and I (indeterminate). The classical law of the excluded third (a statement must be either true or false; there is no third possibility) is replaced by the law of the excluded fourth. Every statement must be true, false, or indeterminate; there is no fourth alternative. For example, statement B, about the momentum of a particle, may be found true if a suitable experiment is made. In that case, the other statement, A, about the particle's position, is indeterminate. It is indeterminate because it is impossible in principle to determine its truth or falsity at the same instant that statement B is confirmed. Of course, A could have been confirmed instead. Then B would have been indeterminate. In other words, there are situations in modern physics in which, if certain statements are true, other statements must be indeterminate.

In order to accommodate his three truth values, Reichenbach found it necessary to redefine the customary logical connectives (implication, disjunction, conjunction, and so on) by truth tables much more complicated than those used to define the connectives of the familiar two-valued logic. In addition, he was led to introduce new connectives. Again, my feeling is that, if it were necessary to complicate logic in this way for the language of physics, it would be acceptable. At present, however, I cannot see the necessity for such a radical step.

We must, of course, wait to see how things go in the future development of physics. Unfortunately, physicists seldom present their theories in a form that logicians would like to see. They do not say: "This is my language, these are the primitive terms, here are my rules of formation, there are the logical axioms." (If they gave at least their logical axioms, we could then see whether they were in agreement with von Neumann or with Reichenbach or whether they preferred to retain the classical two-valued logic.) It would also be good to have the postulates of the entire field of physics stated in a systematic form that would include formal logic. If this were done, it would be easier to determine if there are good reasons to change the underlying logic.

[3] See Hans Reichenbach, *Philosophic Foundations of Quantum Mechanics* (Berkeley: University of California Press, 1944).

Here we touch on deep problems, not yet solved, concerning the language of physics. This language is still, except for its mathematical part, largely a natural language; that is, its rules are learned implicitly in practice and seldom formulated explicitly. Of course, thousands of new terms and phrases peculiar to the language of physics have been adopted, and, in a few cases, special rules have been devised to handle some of these technical terms and symbols. Like the languages of other sciences, the language of physics has steadily increased in exactness and overall efficiency. This trend will certainly continue. At the moment, however, the development of quantum mechanics has not yet been fully reflected in a sharpening of the language of physics.

It is difficult to predict how the language of physics will change. But I am convinced that two tendencies, which have led to great improvements in the language of mathematics during the last half century, will prove equally effective in sharpening and clarifying the language of physics: the application of modern logic and set theory, and the adoption of the axiomatic method in its modern form, which presupposes a formalized language system. In present-day physics, in which not only the content of theories but also the entire conceptual structure of physics is under discussion, both those methods could be of enormous help.

Here is an exciting challenge, which calls for close cooperation between physicists and logicians—better still, for the work of younger men who have studied both physics and logic. The application of modern logic and the axiomatic method to physics will, I believe, do much more than just improve communication among physicists and between physicists and other scientists. It will accomplish something of far greater importance: it will make it easier to create new concepts, to formulate fresh assumptions. An enormous amount of new experimental results has been collected in recent years, much of it due to the great improvement of experimental instruments, such as the big atom smashers. On the basis of these results, great progress has been made in the development of quantum mechanics. Unfortunately, efforts to rebuild the theory, in such a way that all the new data fit into it, have not been successful. Some surprising puzzles and bewildering quandaries have appeared. Their solution is an urgent, but most difficult, task. It seems a fair assumption that the use of new conceptual tools could here be of essential help.

Some physicists believe that there is a good chance for a new breakthrough in the near future. Whether it will be soon or later, we may trust—provided the world's leading statesmen refrain from the ultimate folly of nuclear war and permit humanity to survive—that science will continue to make great progress and lead us to ever deeper insights into the structure of the world.

Bibliography

GENERAL BOOKS

Richard B. Braithwaite. *Scientific Explanation*. Cambridge: Cambridge University Press, 1953.

Percy W. Bridgman. *The Logic of Modern Physics*. New York: Macmillan, 1927.

Norman R. Campbell. *Physics: The Elements*. Cambridge: Cambridge University Press, 1920; reprinted as *Foundations of Science*, New York: Dover, 1957.

Norman R. Campbell. *What Is Science?* London: Methuen, 1921; reprinted New York: Dover, 1953.

Philipp Frank. *Philosophy of Science*. Englewood Cliffs, N.J.: Prentice-Hall, 1957.

Werner Heisenberg. *Physics and Philosophy: The Revolution in Modern Science*. New York: Harper, 1958.

Carl G. Hempel. *Aspects of Scientific Explanation and Other Essays in the Philosophy of Science*. New York: Free Press, 1965.

Carl G. Hempel. *International Encyclopedia of Unified Science*, Vol. 2, No. 7: *Fundamentals of Concept Formation in Empirical Science*. Chicago: University of Chicago Press, 1952.

Gerald Holton and Duane Roller. *Foundations of Modern Physical Science*. Reading, Mass.: Addison-Wesley, 1958.

293

John Kemeny. *A Philosopher Looks at Science.* Princeton, N.J.: D. Van Nostrand, 1959.

Ernest Nagel. *The Structure of Science.* New York: Harcourt, Brace & World, 1961.

Karl Popper. *The Logic of Scientific Discovery.* New York: Basic Books, 1959.

Bertrand Russell. *Human Knowledge: Its Scope and Limits.* New York: Simon & Schuster, 1948.

Israel Scheffler. *The Anatomy of Inquiry.* New York: Knopf, 1963.

Stephen Toulmin. *The Philosophy of Science.* London: Hutchinson, 1953.

COLLECTIONS OF ARTICLES

Arthur Danto and Sidney Morgenbesser, eds. *Philosophy of Science.* New York: Meridian, 1960.

Herbert Feigl and May Brodbeck, eds. *Readings in the Philosophy of Science.* New York: Appleton-Century-Crofts, 1953.

Herbert Feigl and Wilfrid Sellars, eds. *Readings in Philosophical Analysis.* New York: Appleton-Century-Crofts, 1949.

Herbert Feigl, Michael Scriven, and Grover Maxwell, eds. *Minnesota Studies in the Philosophy of Science.* Minneapolis, Minn.: University of Minnesota Press, Vol. I, 1956; Vol. II, 1958; Vol. III, 1962.

Edward H. Madden, ed. *The Structure of Scientific Thought.* Boston, Mass.: Houghton Mifflin, 1960.

Paul Arthur Schilpp, ed. *The Philosophy of Rudolf Carnap.* La Salle, Ill.: Open Court, 1963.

Paul Arthur Schilpp, ed. *Albert Einstein: Philosopher-Scientist.* Evanston, Ill.: Library of Living Philosophers, 1949.

Philip Wiener, ed. *Readings in Philosophy of Science.* New York: Scribner, 1953.

MEASUREMENT

Norman R. Campbell. *Physics: The Elements, op. cit.,* Part II: Measurement.

Carl G. Hempel. *Fundamentals of Concept Formation in Empirical Science, op. cit.,* Chapter 3.

Victor F. Lenzen. *International Encyclopedia of Unified Science,* Vol. I, No. 5: *Procedures of Empirical Science.* Chicago, Ill.: University of Chicago Press, 1938.

SPACE AND TIME

Albert Einstein. *Sidelights on Relativity.* London: Methuen, 1922; reprinted New York: Dover, 1983.

Philipp Frank. *Philosophy of Science, op. cit.,* chapters 3 and 6.

Adolf Grünbaum. *Philosophical Problems of Space and Time.* New York: Knopf, 1963.

Max Jammer. *Concepts of Space.* Cambridge, Mass.: Harvard University Press, 1954, 1969; enlarged edition, New York: Dover, 1993.

Ernest Nagel. *The Structure of Science, op. cit.,* chapters 8 and 9.

Henri Poincaré. *Science and Hypothesis.* London: Walter Scott, 1905; reprinted New York: Dover, 1952.

Hans Reichenbach. *The Philosophy of Space and Time.* New York: Dover, 1958.

THE MEANING OF CAUSALITY

Bertrand Russell. *Mysticism and Logic,* Chapter 9. London: Longmans, Green, 1918.

Bertrand Russell. *Our Knowledge of the External World,* Chapter 8. London: Allen & Unwin, 1914.

Moritz Schlick. "Causality in Everyday Life and in Recent Science." Reprinted in Feigl and Sellars, *Readings in Philosophical Analysis, op. cit.,* pp. 515–533.

DETERMINISM AND FREE WILL

Bertrand Russell. *Our Knowledge of the External World, op. cit.,* Chapter 8.

Moritz Schlick. *Problems of Ethics,* Chapter 7. New York: Prentice-Hall, 1939.

Charles Stevenson. *Ethics and Language.* New Haven: Yale University Press, 1944. Chapter 11.

Index

A CATALOG OF SELECTED
DOVER BOOKS
IN ALL FIELDS OF INTEREST

A CATALOG OF SELECTED DOVER
BOOKS IN ALL FIELDS OF INTEREST

CONCERNING THE SPIRITUAL IN ART, Wassily Kandinsky. Pioneering work by father of abstract art. Thoughts on color theory, nature of art. Analysis of earlier masters. 12 illustrations. 80pp. of text. 5⅜ × 8½.   23411-8 Pa. $3.95

ANIMALS: 1,419 Copyright-Free Illustrations of Mammals, Birds, Fish, Insects, etc., Jim Harter (ed.). Clear wood engravings present, in extremely lifelike poses, over 1,000 species of animals. One of the most extensive pictorial sourcebooks of its kind. Captions. Index. 284pp. 9 × 12.   23766-4 Pa. $11.95

CELTIC ART: The Methods of Construction, George Bain. Simple geometric techniques for making Celtic interlacements, spirals, Kells-type initials, animals, humans, etc. Over 500 illustrations. 160pp. 9 × 12. (USO)   22923-8 Pa. $8.95

AN ATLAS OF ANATOMY FOR ARTISTS, Fritz Schider. Most thorough reference work on art anatomy in the world. Hundreds of illustrations, including selections from works by Vesalius, Leonardo, Goya, Ingres, Michelangelo, others. 593 illustrations. 192pp. 7⅛ × 10¼.   20241-0 Pa. $8.95

CELTIC HAND STROKE-BY-STROKE (Irish Half-Uncial from "The Book of Kells"): An Arthur Baker Calligraphy Manual, Arthur Baker. Complete guide to creating each letter of the alphabet in distinctive Celtic manner. Covers hand position, strokes, pens, inks, paper, more. Illustrated. 48pp. 8¼ × 11.
24336-2 Pa. $3.95

EASY ORIGAMI, John Montroll. Charming collection of 32 projects (hat, cup, pelican, piano, swan, many more) specially designed for the novice origami hobbyist. Clearly illustrated easy-to-follow instructions insure that even beginning papercrafters will achieve successful results. 48pp. 8¼ × 11.   27298-2 Pa. $2.95

THE COMPLETE BOOK OF BIRDHOUSE CONSTRUCTION FOR WOOD-WORKERS, Scott D. Campbell. Detailed instructions, illustrations, tables. Also data on bird habitat and instinct patterns. Bibliography. 3 tables. 63 illustrations in 15 figures. 48pp. 5¼ × 8½.   24407-5 Pa. $1.95

BLOOMINGDALE'S ILLUSTRATED 1886 CATALOG: Fashions, Dry Goods and Housewares, Bloomingdale Brothers. Famed merchants' extremely rare catalog depicting about 1,700 products: clothing, housewares, firearms, dry goods, jewelry, more. Invaluable for dating, identifying vintage items. Also, copyright-free graphics for artists, designers. Co-published with Henry Ford Museum & Green-field Village. 160pp. 8¼ × 11.   25780-0 Pa. $9.95

HISTORIC COSTUME IN PICTURES, Braun & Schneider. Over 1,450 costumed figures in clearly detailed engravings—from dawn of civilization to end of 19th century. Captions. Many folk costumes. 256pp. 8⅜ × 11¾.   23150-X Pa. $10.95

STICKLEY CRAFTSMAN FURNITURE CATALOGS, Gustav Stickley and L. & J. G. Stickley. Beautiful, functional furniture in two authentic catalogs from 1910. 594 illustrations, including 277 photos, show settles, rockers, armchairs, reclining chairs, bookcases, desks, tables. 183pp. 6½ × 9¼. 23838-5 Pa. $8.95

AMERICAN LOCOMOTIVES IN HISTORIC PHOTOGRAPHS: 1858 to 1949, Ron Ziel (ed.). A rare collection of 126 meticulously detailed official photographs, called "builder portraits," of American locomotives that majestically chronicle the rise of steam locomotive power in America. Introduction. Detailed captions. xi + 129pp. 9 × 12. 27393-8 Pa. $12.95

AMERICA'S LIGHTHOUSES: An Illustrated History, Francis Ross Holland, Jr. Delightfully written, profusely illustrated fact-filled survey of over 200 American lighthouses since 1716. History, anecdotes, technological advances, more. 240pp. 8 × 10¾. 25576-X Pa. $11.95

TOWARDS A NEW ARCHITECTURE, Le Corbusier. Pioneering manifesto by founder of "International School." Technical and aesthetic theories, views of industry, economics, relation of form to function, "mass-production split" and much more. Profusely illustrated. 320pp. 6⅛ × 9¼. (USO) 25023-7 Pa. $8.95

HOW THE OTHER HALF LIVES, Jacob Riis. Famous journalistic record, exposing poverty and degradation of New York slums around 1900, by major social reformer. 100 striking and influential photographs. 233pp. 10 × 7⅞.
22012-5 Pa $10.95

FRUIT KEY AND TWIG KEY TO TREES AND SHRUBS, William M. Harlow. One of the handiest and most widely used identification aids. Fruit key covers 120 deciduous and evergreen species; twig key 160 deciduous species. Easily used. Over 300 photographs. 126pp. 5⅜ × 8½. 20511-8 Pa. $3.95

COMMON BIRD SONGS, Dr. Donald J. Borror. Songs of 60 most common U.S. birds: robins, sparrows, cardinals, bluejays, finches, more—arranged in order of increasing complexity. Up to 9 variations of songs of each species.
Cassette and manual 99911-4 $8.95

ORCHIDS AS HOUSE PLANTS, Rebecca Tyson Northen. Grow cattleyas and many other kinds of orchids—in a window, in a case, or under artificial light. 63 illustrations. 148pp. 5⅜ × 8½. 23261-1 Pa. $3.95

MONSTER MAZES, Dave Phillips. Masterful mazes at four levels of difficulty. Avoid deadly perils and evil creatures to find magical treasures. Solutions for all 32 exciting illustrated puzzles. 48pp. 8¼ × 11. 26005-4 Pa. $2.95

MOZART'S DON GIOVANNI (DOVER OPERA LIBRETTO SERIES), Wolfgang Amadeus Mozart. Introduced and translated by Ellen H. Bleiler. Standard Italian libretto, with complete English translation. Convenient and thoroughly portable—an ideal companion for reading along with a recording or the performance itself. Introduction. List of characters. Plot summary. 121pp. 5¼ × 8½.
24944-1 Pa. $2.95

TECHNICAL MANUAL AND DICTIONARY OF CLASSICAL BALLET, Gail Grant. Defines, explains, comments on steps, movements, poses and concepts. 15-page pictorial section. Basic book for student, viewer. 127pp. 5⅜ × 8½.
21843-0 Pa. $3.95

CATALOG OF DOVER BOOKS

BRASS INSTRUMENTS: Their History and Development, Anthony Baines. Authoritative, updated survey of the evolution of trumpets, trombones, bugles, cornets, French horns, tubas and other brass wind instruments. Over 140 illustrations and 48 music examples. Corrected and updated by author. New preface. Bibliography. 320pp. 5⅜ × 8½. 27574-4 Pa. $9.95

HOLLYWOOD GLAMOR PORTRAITS, John Kobal (ed.). 145 photos from 1926–49. Harlow, Gable, Bogart, Bacall; 94 stars in all. Full background on photographers, technical aspects. 160pp. 8⅜ × 11¼. 23352-9 Pa. $9.95

MAX AND MORITZ, Wilhelm Busch. Great humor classic in both German and English. Also 10 other works: "Cat and Mouse," "Plisch and Plumm," etc. 216pp. 5⅜ × 8½. 20181-3 Pa. $5.95

THE RAVEN AND OTHER FAVORITE POEMS, Edgar Allan Poe. Over 40 of the author's most memorable poems: "The Bells," "Ulalume," "Israfel," "To Helen," "The Conqueror Worm," "Eldorado," "Annabel Lee," many more. Alphabetic lists of titles and first lines. 64pp. 5³⁄₁₆ × 8¼. 26685-0 Pa. $1.00

SEVEN SCIENCE FICTION NOVELS, H. G. Wells. The standard collection of the great novels. Complete, unabridged. First Men in the Moon, Island of Dr. Moreau, War of the Worlds, Food of the Gods, Invisible Man, Time Machine, In the Days of the Comet. Total of 1,015pp. 5⅜ × 8½. (USO) 20264-X Clothbd. $29.95

AMULETS AND SUPERSTITIONS, E. A. Wallis Budge. Comprehensive discourse on origin, powers of amulets in many ancient cultures: Arab, Persian, Babylonian, Assyrian, Egyptian, Gnostic, Hebrew, Phoenician, Syriac, etc. Covers cross, swastika, crucifix, seals, rings, stones, etc. 584pp. 5⅜ × 8½. 23573-4 Pa. $12.95

RUSSIAN STORIES/PYCCKNE PACCKA3bl: A Dual-Language Book, edited by Gleb Struve. Twelve tales by such masters as Chekhov, Tolstoy, Dostoevsky, Pushkin, others. Excellent word-for-word English translations on facing pages, plus teaching and study aids, Russian/English vocabulary, biographical/critical introductions, more. 416pp. 5⅜ × 8½. 26244-8 Pa. $8.95

PHILADELPHIA THEN AND NOW: 60 Sites Photographed in the Past and Present, Kenneth Finkel and Susan Oyama. Rare photographs of City Hall, Logan Square, Independence Hall, Betsy Ross House, other landmarks juxtaposed with contemporary views. Captures changing face of historic city. Introduction. Captions. 128pp. 8¼ × 11. 25790-8 Pa. $9.95

AIA ARCHITECTURAL GUIDE TO NASSAU AND SUFFOLK COUNTIES, LONG ISLAND, The American Institute of Architects, Long Island Chapter, and the Society for the Preservation of Long Island Antiquities. Comprehensive, well-researched and generously illustrated volume brings to life over three centuries of Long Island's great architectural heritage. More than 240 photographs with authoritative, extensively detailed captions. 176pp. 8¼ × 11. 26946-9 Pa. $14.95

NORTH AMERICAN INDIAN LIFE: Customs and Traditions of 23 Tribes, Elsie Clews Parsons (ed.). 27 fictionalized essays by noted anthropologists examine religion, customs, government, additional facets of life among the Winnebago, Crow, Zuni, Eskimo, other tribes. 480pp. 6⅛ × 9¼. 27377-6 Pa. $10.95

FRANK LLOYD WRIGHT'S HOLLYHOCK HOUSE, Donald Hoffmann. Lavishly illustrated, carefully documented study of one of Wright's most controversial residential designs. Over 120 photographs, floor plans, elevations, etc. Detailed perceptive text by noted Wright scholar. Index. 128pp. 9¼ × 10¾.
27133-1 Pa. $11.95

THE MALE AND FEMALE FIGURE IN MOTION: 60 Classic Photographic Sequences, Eadweard Muybridge. 60 true-action photographs of men and women walking, running, climbing, bending, turning, etc., reproduced from rare 19th-century masterpiece. vi + 121pp. 9 × 12. 24745-7 Pa. $10.95

1001 QUESTIONS ANSWERED ABOUT THE SEASHORE, N. J. Berrill and Jacquelyn Berrill. Queries answered about dolphins, sea snails, sponges, starfish, fishes, shore birds, many others. Covers appearance, breeding, growth, feeding, much more. 305pp. 5¼ × 8¼. 23366-9 Pa. $7.95

GUIDE TO OWL WATCHING IN NORTH AMERICA, Donald S. Heintzelman. Superb guide offers complete data and descriptions of 19 species: barn owl, screech owl, snowy owl, many more. Expert coverage of owl-watching equipment, conservation, migrations and invasions, etc. Guide to observing sites. 84 illustrations. xiii + 193pp. 5⅜ × 8½. 27344-X Pa. $7.95

MEDICINAL AND OTHER USES OF NORTH AMERICAN PLANTS: A Historical Survey with Special Reference to the Eastern Indian Tribes, Charlotte Erichsen-Brown. Chronological historical citations document 500 years of usage of plants, trees, shrubs native to eastern Canada, northeastern U.S. Also complete identifying information. 343 illustrations. 544pp. 6½ × 9¼. 25951-X Pa. $12.95

STORYBOOK MAZES, Dave Phillips. 23 stories and mazes on two-page spreads: Wizard of Oz, Treasure Island, Robin Hood, etc. Solutions. 64pp. 8¼ × 11. 23628-5 Pa. $2.95

NEGRO FOLK MUSIC, U.S.A., Harold Courlander. Noted folklorist's scholarly yet readable analysis of rich and varied musical tradition. Includes authentic versions of over 40 folk songs. Valuable bibliography and discography. xi + 324pp. 5⅜ × 8½. 27350-4 Pa. $7.95

MOVIE-STAR PORTRAITS OF THE FORTIES, John Kobal (ed.). 163 glamor, studio photos of 106 stars of the 1940s: Rita Hayworth, Ava Gardner, Marlon Brando, Clark Gable, many more. 176pp. 8⅜ × 11¼. 23546-7 Pa. $10.95

BENCHLEY LOST AND FOUND, Robert Benchley. Finest humor from early 30s, about pet peeves, child psychologists, post office and others. Mostly unavailable elsewhere. 73 illustrations by Peter Arno and others. 183pp. 5⅜ × 8½. 22410-4 Pa. $5.95

YEKL and THE IMPORTED BRIDEGROOM AND OTHER STORIES OF YIDDISH NEW YORK, Abraham Cahan. Film Hester Street based on Yekl (1896). Novel, other stories among first about Jewish immigrants on N.Y.'s East Side. 240pp. 5⅜ × 8½. 22427-9 Pa. $5.95

SELECTED POEMS, Walt Whitman. Generous sampling from *Leaves of Grass*. Twenty-four poems include "I Hear America Singing," "Song of the Open Road," "I Sing the Body Electric," "When Lilacs Last in the Dooryard Bloom'd," "O Captain! My Captain!"—all reprinted from an authoritative edition. Lists of titles and first lines. 128pp. 5³⁄₁₆ × 8¼. 26878-0 Pa. $1.00

THE BEST TALES OF HOFFMANN, E. T. A. Hoffmann. 10 of Hoffmann's most important stories: "Nutcracker and the King of Mice," "The Golden Flowerpot," etc. 458pp. 5⅜ × 8½. 21793-0 Pa. $8.95

FROM FETISH TO GOD IN ANCIENT EGYPT, E. A. Wallis Budge. Rich detailed survey of Egyptian conception of "God" and gods, magic, cult of animals, Osiris, more. Also, superb English translations of hymns and legends. 240 illustrations. 545pp. 5⅜ × 8½. 25803-3 Pa. $11.95

FRENCH STORIES/CONTES FRANÇAIS: A Dual-Language Book, Wallace Fowlie. Ten stories by French masters, Voltaire to Camus: "Micromegas" by Voltaire; "The Atheist's Mass" by Balzac; "Minuet" by de Maupassant; "The Guest" by Camus, six more. Excellent English translations on facing pages. Also French-English vocabulary list, exercises, more. 352pp. 5⅜ × 8½. 26443-2 Pa. $8.95

CHICAGO AT THE TURN OF THE CENTURY IN PHOTOGRAPHS: 122 Historic Views from the Collections of the Chicago Historical Society, Larry A. Viskochil. Rare large-format prints offer detailed views of City Hall, State Street, the Loop, Hull House, Union Station, many other landmarks, circa 1904–1913. Introduction. Captions. Maps. 144pp. 9⅜ × 12¼. 24656-6 Pa. $12.95

OLD BROOKLYN IN EARLY PHOTOGRAPHS, 1865–1929, William Lee Younger. Luna Park, Gravesend race track, construction of Grand Army Plaza, moving of Hotel Brighton, etc. 157 previously unpublished photographs. 165pp. 8⅞ × 11¾. 23587-4 Pa. $12.95

THE MYTHS OF THE NORTH AMERICAN INDIANS, Lewis Spence. Rich anthology of the myths and legends of the Algonquins, Iroquois, Pawnees and Sioux, prefaced by an extensive historical and ethnological commentary. 36 illustrations. 480pp. 5⅜ × 8½. 25967-6 Pa. $8.95

AN ENCYCLOPEDIA OF BATTLES: Accounts of Over 1,560 Battles from 1479 B.C. to the Present, David Eggenberger. Essential details of every major battle in recorded history from the first battle of Megiddo in 1479 B.C. to Grenada in 1984. List of Battle Maps. New Appendix covering the years 1967–1984. Index. 99 illustrations. 544pp. 6½ × 9¼. 24913-1 Pa. $14.95

SAILING ALONE AROUND THE WORLD, Captain Joshua Slocum. First man to sail around the world, alone, in small boat. One of great feats of seamanship told in delightful manner. 67 illustrations. 294pp. 5⅜ × 8½. 20326-3 Pa. $5.95

ANARCHISM AND OTHER ESSAYS, Emma Goldman. Powerful, penetrating, prophetic essays on direct action, role of minorities, prison reform, puritan hypocrisy, violence, etc. 271pp. 5⅜ × 8½. 22484-8 Pa. $5.95

MYTHS OF THE HINDUS AND BUDDHISTS, Ananda K. Coomaraswamy and Sister Nivedita. Great stories of the epics; deeds of Krishna, Shiva, taken from puranas, Vedas, folk tales; etc. 32 illustrations. 400pp. 5⅜ × 8½. 21759-0 Pa. $9.95

BEYOND PSYCHOLOGY, Otto Rank. Fear of death, desire of immortality, nature of sexuality, social organization, creativity, according to Rankian system. 291pp. 5⅜ × 8½. 20485-5 Pa. $7.95

A THEOLOGICO-POLITICAL TREATISE, Benedict Spinoza. Also contains unfinished Political Treatise. Great classic on religious liberty, theory of government on common consent. R. Elwes translation. Total of 421pp. 5⅜ × 8½. 20249-6 Pa. $7.95

CATALOG OF DOVER BOOKS

MY BONDAGE AND MY FREEDOM, Frederick Douglass. Born a slave, Douglass became outspoken force in antislavery movement. The best of Douglass' autobiographies. Graphic description of slave life. 464pp. 5⅜ × 8½. 22457-0 Pa. $8.95

FOLLOWING THE EQUATOR: A Journey Around the World, Mark Twain. Fascinating humorous account of 1897 voyage to Hawaii, Australia, India, New Zealand, etc. Ironic, bemused reports on peoples, customs, climate, flora and fauna, politics, much more. 197 illustrations. 720pp. 5⅜ × 8½. 26113-1 Pa. $15.95

THE PEOPLE CALLED SHAKERS, Edward D. Andrews. Definitive study of Shakers: origins, beliefs, practices, dances, social organization, furniture and crafts, etc. 33 illustrations. 351pp. 5⅜ × 8½. 21081-2 Pa. $7.95

THE MYTHS OF GREECE AND ROME, H. A. Guerber. A classic of mythology, generously illustrated, long prized for its simple, graphic, accurate retelling of the principal myths of Greece and Rome, and for its commentary on their origins and significance. With 64 illustrations by Michelangelo, Raphael, Titian, Rubens, Canova, Bernini and others. 480pp. 5⅜ × 8½. 27584-1 Pa. $9.95

PSYCHOLOGY OF MUSIC, Carl E. Seashore. Classic work discusses music as a medium from psychological viewpoint. Clear treatment of physical acoustics, auditory apparatus, sound perception, development of musical skills, nature of musical feeling, host of other topics. 88 figures. 408pp. 5⅜ × 8½. 21851-1 Pa. $9.95

THE PHILOSOPHY OF HISTORY, Georg W. Hegel. Great classic of Western thought develops concept that history is not chance but rational process, the evolution of freedom. 457pp. 5⅜ × 8½. 20112-0 Pa. $8.95

THE BOOK OF TEA, Kakuzo Okakura. Minor classic of the Orient: entertaining, charming explanation, interpretation of traditional Japanese culture in terms of tea ceremony. 94pp. 5⅜ × 8½. 20070-1 Pa. $2.95

LIFE IN ANCIENT EGYPT, Adolf Erman. Fullest, most thorough, detailed older account with much not in more recent books, domestic life, religion, magic, medicine, commerce, much more. Many illustrations reproduce tomb paintings, carvings, hieroglyphs, etc. 597pp. 5⅜ × 8½. 22632-8 Pa. $9.95

SUNDIALS, Their Theory and Construction, Albert Waugh. Far and away the best, most thorough coverage of ideas, mathematics concerned, types, construction, adjusting anywhere. Simple, nontechnical treatment allows even children to build several of these dials. Over 100 illustrations. 230pp. 5⅜ × 8½. 22947-5 Pa. $5.95

DYNAMICS OF FLUIDS IN POROUS MEDIA, Jacob Bear. For advanced students of ground water hydrology, soil mechanics and physics, drainage and irrigation engineering, and more. 335 illustrations. Exercises, with answers. 784pp. 6⅛ × 9¼. 65675-6 Pa. $19.95

SONGS OF EXPERIENCE: Facsimile Reproduction with 26 Plates in Full Color, William Blake. 26 full-color plates from a rare 1826 edition. Includes "The Tyger," "London," "Holy Thursday," and other poems. Printed text of poems. 48pp. 5¼ × 7. 24636-1 Pa. $3.95

OLD-TIME VIGNETTES IN FULL COLOR, Carol Belanger Grafton (ed.). Over 390 charming, often sentimental illustrations, selected from archives of Victorian graphics—pretty women posing, children playing, food, flowers, kittens and puppies, smiling cherubs, birds and butterflies, much more. All copyright-free. 48pp. 9¼ × 12¼. 27269-9 Pa. $5.95

PERSPECTIVE FOR ARTISTS, Rex Vicat Cole. Depth, perspective of sky and sea, shadows, much more, not usually covered. 391 diagrams, 81 reproductions of drawings and paintings. 279pp. 5⅜ × 8½. 22487-2 Pa. $6.95

DRAWING THE LIVING FIGURE, Joseph Sheppard. Innovative approach to artistic anatomy focuses on specifics of surface anatomy, rather than muscles and bones. Over 170 drawings of live models in front, back and side views, and in widely varying poses. Accompanying diagrams. 177 illustrations. Introduction. Index. 144pp. 8⅜ × 11¼. 26723-7 Pa. $7.95

GOTHIC AND OLD ENGLISH ALPHABETS: 100 Complete Fonts, Dan X. Solo. Add power, elegance to posters, signs, other graphics with 100 stunning copyright-free alphabets: Blackstone, Dolbey, Germania, 97 more—including many lower-case, numerals, punctuation marks. 104pp. 8⅛ × 11. 24695-7 Pa. $7.95

HOW TO DO BEADWORK, Mary White. Fundamental book on craft from simple projects to five-bead chains and woven works. 106 illustrations. 142pp. 5⅜ × 8. 20697-1 Pa. $4.95

THE BOOK OF WOOD CARVING, Charles Marshall Sayers. Finest book for beginners discusses fundamentals and offers 34 designs. "Absolutely first rate . . . well thought out and well executed."—E. J. Tangerman. 118pp. 7¾ × 10⅝. 23654-4 Pa. $5.95

ILLUSTRATED CATALOG OF CIVIL WAR MILITARY GOODS: Union Army Weapons, Insignia, Uniform Accessories, and Other Equipment, Schuyler, Hartley, and Graham. Rare, profusely illustrated 1846 catalog includes Union Army uniform and dress regulations, arms and ammunition, coats, insignia, flags, swords, rifles, etc. 226 illustrations. 160pp. 9 × 12. 24939-5 Pa. $10.95

WOMEN'S FASHIONS OF THE EARLY 1900s: An Unabridged Republication of "New York Fashions, 1909," National Cloak & Suit Co. Rare catalog of mail-order fashions documents women's and children's clothing styles shortly after the turn of the century. Captions offer full descriptions, prices. Invaluable resource for fashion, costume historians. Approximately 725 illustrations. 128pp. 8⅜ × 11¼. 27276-1 Pa. $10.95

THE 1912 AND 1915 GUSTAV STICKLEY FURNITURE CATALOGS, Gustav Stickley. With over 200 detailed illustrations and descriptions, these two catalogs are essential reading and reference materials and identification guides for Stickley furniture. Captions cite materials, dimensions and prices. 112pp. 6½ × 9¼. 26676-1 Pa. $9.95

EARLY AMERICAN LOCOMOTIVES, John H. White, Jr. Finest locomotive engravings from early 19th century: historical (1804–74), main-line (after 1870), special, foreign, etc. 147 plates. 142pp. 11⅜ × 8¼. 22772-3 Pa. $8.95

THE TALL SHIPS OF TODAY IN PHOTOGRAPHS, Frank O. Braynard. Lavishly illustrated tribute to nearly 100 majestic contemporary sailing vessels: Amerigo Vespucci, Clearwater, Constitution, Eagle, Mayflower, Sea Cloud, Victory, many more. Authoritative captions provide statistics, background on each ship. 190 black-and-white photographs and illustrations. Introduction. 128pp. 8⅞ × 11¾. 27163-3 Pa. $12.95

EARLY NINETEENTH-CENTURY CRAFTS AND TRADES, Peter Stockham (ed.). Extremely rare 1807 volume describes to youngsters the crafts and trades of the day: brickmaker, weaver, dressmaker, bookbinder, ropemaker, saddler, many more. Quaint prose, charming illustrations for each craft. 20 black-and-white line illustrations. 192pp. 4⅝ × 6. 27293-1 Pa. $4.95

VICTORIAN FASHIONS AND COSTUMES FROM HARPER'S BAZAR, 1867-1898, Stella Blum (ed.). Day costumes, evening wear, sports clothes, shoes, hats, other accessories in over 1,000 detailed engravings. 320pp. 9⅜ × 12¼. 22990-4 Pa. $13.95

GUSTAV STICKLEY, THE CRAFTSMAN, Mary Ann Smith. Superb study surveys broad scope of Stickley's achievement, especially in architecture. Design philosophy, rise and fall of the Craftsman empire, descriptions and floor plans for many Craftsman houses, more. 86 black-and-white halftones. 31 line illustrations. Introduction. 208pp. 6½ × 9¼. 27210-9 Pa. $9.95

THE LONG ISLAND RAIL ROAD IN EARLY PHOTOGRAPHS, Ron Ziel. Over 220 rare photos, informative text document origin (1844) and development of rail service on Long Island. Vintage views of early trains, locomotives, stations, passengers, crews, much more. Captions. 8⅞ × 11¾. 26301-0 Pa. $13.95

THE BOOK OF OLD SHIPS: From Egyptian Galleys to Clipper Ships, Henry B. Culver. Superb, authoritative history of sailing vessels, with 80 magnificent line illustrations. Galley, bark, caravel, longship, whaler, many more. Detailed, informative text on each vessel by noted naval historian. Introduction. 256pp. 5⅜ × 8½. 27332-6 Pa. $6.95

TEN BOOKS ON ARCHITECTURE, Vitruvius. The most important book ever written on architecture. Early Roman aesthetics, technology, classical orders, site selection, all other aspects. Morgan translation. 331pp. 5⅜ × 8½. 20645-9 Pa. $8.95

THE HUMAN FIGURE IN MOTION, Eadweard Muybridge. More than 4,500 stopped-action photos, in action series, showing undraped men, women, children jumping, lying down, throwing, sitting, wrestling, carrying, etc. 390pp. 7⅞ × 10⅝. 20204-6 Clothbd. $24.95

TREES OF THE EASTERN AND CENTRAL UNITED STATES AND CANADA, William M. Harlow. Best one-volume guide to 140 trees. Full descriptions, woodlore, range, etc. Over 600 illustrations. Handy size. 288pp. 4½ × 6⅜. 20395-6 Pa. $5.95

SONGS OF WESTERN BIRDS, Dr. Donald J. Borror. Complete song and call repertoire of 60 western species, including flycatchers, juncoes, cactus wrens, many more—includes fully illustrated booklet. Cassette and manual 99913-0 $8.95

GROWING AND USING HERBS AND SPICES, Milo Miloradovich. Versatile handbook provides all the information needed for cultivation and use of all the herbs and spices available in North America. 4 illustrations. Index. Glossary. 236pp. 5⅜ × 8½. 25058-X Pa. $5.95

BIG BOOK OF MAZES AND LABYRINTHS, Walter Shepherd. 50 mazes and labyrinths in all—classical, solid, ripple, and more—in one great volume. Perfect inexpensive puzzler for clever youngsters. Full solutions. 112pp. 8⅛ × 11. 22951-3 Pa. $3.95

CATALOG OF DOVER BOOKS

PIANO TUNING, J. Cree Fischer. Clearest, best book for beginner, amateur. Simple repairs, raising dropped notes, tuning by easy method of flattened fifths. No previous skills needed. 4 illustrations. 201pp. 5⅜ × 8½. 23267-0 Pa. $5.95

A SOURCE BOOK IN THEATRICAL HISTORY, A. M. Nagler. Contemporary observers on acting, directing, make-up, costuming, stage props, machinery, scene design, from Ancient Greece to Chekhov. 611pp. 5⅜ × 8½. 20515-0 Pa. $11.95

THE COMPLETE NONSENSE OF EDWARD LEAR, Edward Lear. All nonsense limericks, zany alphabets, Owl and Pussycat, songs, nonsense botany, etc., illustrated by Lear. Total of 320pp. 5⅜ × 8½. (USO) 20167-8 Pa. $5.95

VICTORIAN PARLOUR POETRY: An Annotated Anthology, Michael R. Turner. 117 gems by Longfellow, Tennyson, Browning, many lesser-known poets. "The Village Blacksmith," "Curfew Must Not Ring Tonight," "Only a Baby Small," dozens more, often difficult to find elsewhere. Index of poets, titles, first lines. xxiii + 325pp. 5⅜ × 8¼. 27044-0 Pa. $8.95

DUBLINERS, James Joyce. Fifteen stories offer vivid, tightly focused observations of the lives of Dublin's poorer classes. At least one, "The Dead," is considered a masterpiece. Reprinted complete and unabridged from standard edition. 160pp. 5³⁄₁₆ × 8¼. 26870-5 Pa. $1.00

THE HAUNTED MONASTERY and THE CHINESE MAZE MURDERS, Robert van Gulik. Two full novels by van Gulik, set in 7th-century China, continue adventures of Judge Dee and his companions. An evil Taoist monastery, seemingly supernatural events; overgrown topiary maze hides strange crimes. 27 illustrations. 328pp. 5⅜ × 8½. 23502-5 Pa. $7.95

THE BOOK OF THE SACRED MAGIC OF ABRAMELIN THE MAGE, translated by S. MacGregor Mathers. Medieval manuscript of ceremonial magic. Basic document in Aleister Crowley, Golden Dawn groups. 268pp. 5⅜ × 8½. 23211-5 Pa. $7.95

NEW RUSSIAN-ENGLISH AND ENGLISH-RUSSIAN DICTIONARY, M. A. O'Brien. This is a remarkably handy Russian dictionary, containing a surprising amount of information, including over 70,000 entries. 366pp. 4½ × 6⅛. 20208-9 Pa. $8.95

HISTORIC HOMES OF THE AMERICAN PRESIDENTS, Second, Revised Edition, Irvin Haas. A traveler's guide to American Presidential homes, most open to the public, depicting and describing homes occupied by every American President from George Washington to George Bush. With visiting hours, admission charges, travel routes. 175 photographs. Index. 160pp. 8¼ × 11. 26751-2 Pa. $10.95

NEW YORK IN THE FORTIES, Andreas Feininger. 162 brilliant photographs by the well-known photographer, formerly with *Life* magazine. Commuters, shoppers, Times Square at night, much else from city at its peak. Captions by John von Hartz. 181pp. 9¼ × 10¾. 23585-8 Pa. $12.95

INDIAN SIGN LANGUAGE, William Tomkins. Over 525 signs developed by Sioux and other tribes. Written instructions and diagrams. Also 290 pictographs. 111pp. 6⅛ × 9¼. 22029-X Pa. $3.50

ANATOMY: A Complete Guide for Artists, Joseph Sheppard. A master of figure drawing shows artists how to render human anatomy convincingly. Over 460 illustrations. 224pp. 8⅜ × 11¼. 27279-6 Pa. $9.95

MEDIEVAL CALLIGRAPHY: Its History and Technique, Marc Drogin. Spirited history, comprehensive instruction manual covers 13 styles (ca. 4th century thru 15th). Excellent photographs; directions for duplicating medieval techniques with modern tools. 224pp. 8⅜ × 11¼. 26142-5 Pa. $11.95

DRIED FLOWERS: How to Prepare Them, Sarah Whitlock and Martha Rankin. Complete instructions on how to use silica gel, meal and borax, perlite aggregate, sand and borax, glycerine and water to create attractive permanent flower arrangements. 12 illustrations. 32pp. 5⅜ × 8½. 21802-3 Pa. $1.00

EASY-TO-MAKE BIRD FEEDERS FOR WOODWORKERS, Scott D. Campbell. Detailed, simple-to-use guide for designing, constructing, caring for and using feeders. Text, illustrations for 12 classic and contemporary designs. 96pp. 5⅜ × 8½. 25847-5 Pa. $2.95

OLD-TIME CRAFTS AND TRADES, Peter Stockham. An 1807 book created to teach children about crafts and trades open to them as future careers. It describes in detailed, nontechnical terms 24 different occupations, among them coachmaker, gardener, hairdresser, lacemaker, shoemaker, wheelwright, copper-plate printer, milliner, trunkmaker, merchant and brewer. Finely detailed engravings illustrate each occupation. 192pp. 4⅝ × 6. 27398-9 Pa. $4.95

THE HISTORY OF UNDERCLOTHES, C. Willett Cunnington and Phyllis Cunnington. Fascinating, well-documented survey covering six centuries of English undergarments, enhanced with over 100 illustrations: 12th-century laced-up bodice, footed long drawers (1795), 19th-century bustles, 19th-century corsets for men, Victorian "bust improvers," much more. 272pp. 5⅜ × 8¼. 27124-2 Pa. $9.95

ARTS AND CRAFTS FURNITURE: The Complete Brooks Catalog of 1912, Brooks Manufacturing Co. Photos and detailed descriptions of more than 150 now very collectible furniture designs from the Arts and Crafts movement depict davenports, settees, buffets, desks, tables, chairs, bedsteads, dressers and more, all built of solid, quarter-sawed oak. Invaluable for students and enthusiasts of antiques, Americana and the decorative arts. 80pp. 6½ × 9¼. 27471-3 Pa. $7.95

HOW WE INVENTED THE AIRPLANE: An Illustrated History, Orville Wright. Fascinating firsthand account covers early experiments, construction of planes and motors, first flights, much more. Introduction and commentary by Fred C. Kelly. 76 photographs. 96pp. 8¼ × 11. 25662-6 Pa. $7.95

THE ARTS OF THE SAILOR: Knotting, Splicing and Ropework, Hervey Garrett Smith. Indispensable shipboard reference covers tools, basic knots and useful hitches; handsewing and canvas work, more. Over 100 illustrations. Delightful reading for sea lovers. 256pp. 5⅜ × 8½. 26440-8 Pa. $7.95

FRANK LLOYD WRIGHT'S FALLINGWATER: The House and Its History, Second, Revised Edition, Donald Hoffmann. A total revision—both in text and illustrations—of the standard document on Fallingwater, the boldest, most personal architectural statement of Wright's mature years, updated with valuable new material from the recently opened Frank Lloyd Wright Archives. "Fascinating"—*The New York Times*. 116 illustrations. 128pp. 9¼ × 10¾. 27430-6 Pa. $10.95

CATALOG OF DOVER BOOKS

PHOTOGRAPHIC SKETCHBOOK OF THE CIVIL WAR, Alexander Gardner. 100 photos taken on field during the Civil War. Famous shots of Manassas, Harper's Ferry, Lincoln, Richmond, slave pens, etc. 244pp. 10⅝ × 8¼.
22731-6 Pa. $9.95

FIVE ACRES AND INDEPENDENCE, Maurice G. Kains. Great back-to-the-land classic explains basics of self-sufficient farming. The one book to get. 95 illustrations. 397pp. 5⅜ × 8½. 20974-1 Pa. $6.95

SONGS OF EASTERN BIRDS, Dr. Donald J. Borror. Songs and calls of 60 species most common to eastern U.S.: warblers, woodpeckers, flycatchers, thrushes, larks, many more in high-quality recording. Cassette and manual 99912-2 $8.95

A MODERN HERBAL, Margaret Grieve. Much the fullest, most exact, most useful compilation of herbal material. Gigantic alphabetical encyclopedia, from aconite to zedoary, gives botanical information, medical properties, folklore, economic uses, much else. Indispensable to serious reader. 161 illustrations. 888pp. 6½ × 9¼. 2-vol. set. (USO) Vol. I: 22798-7 Pa. $9.95
Vol. II: 22799-5 Pa. $9.95

HIDDEN TREASURE MAZE BOOK, Dave Phillips. Solve 34 challenging mazes accompanied by heroic tales of adventure. Evil dragons, people-eating plants, bloodthirsty giants, many more dangerous adversaries lurk at every twist and turn. 34 mazes, stories, solutions. 48pp. 8¼ × 11. 24566-7 Pa. $2.95

LETTERS OF W. A. MOZART, Wolfgang A. Mozart. Remarkable letters show bawdy wit, humor, imagination, musical insights, contemporary musical world; includes some letters from Leopold Mozart. 276pp. 5⅜ × 8½. 22859-2 Pa. $6.95

BASIC PRINCIPLES OF CLASSICAL BALLET, Agrippina Vaganova. Great Russian theoretician, teacher explains methods for teaching classical ballet. 118 illustrations. 175pp. 5⅜ × 8½. 22036-2 Pa. $4.95

THE JUMPING FROG, Mark Twain. Revenge edition. The original story of The Celebrated Jumping Frog of Calaveras County, a hapless French translation, and Twain's hilarious "retranslation" from the French. 12 illustrations. 66pp. 5⅜ × 8½.
22686-7 Pa. $3.50

BEST REMEMBERED POEMS, Martin Gardner (ed.). The 126 poems in this superb collection of 19th- and 20th-century British and American verse range from Shelley's "To a Skylark" to the impassioned "Renascence" of Edna St. Vincent Millay and to Edward Lear's whimsical "The Owl and the Pussycat." 224pp. 5⅜ × 8½.
27165-X Pa. $4.95

COMPLETE SONNETS, William Shakespeare. Over 150 exquisite poems deal with love, friendship, the tyranny of time, beauty's evanescence, death and other themes in language of remarkable power, precision and beauty. Glossary of archaic terms. 80pp. 5³⁄₁₆ × 8¼. 26686-9 Pa. $1.00

BODIES IN A BOOKSHOP, R. T. Campbell. Challenging mystery of blackmail and murder with ingenious plot and superbly drawn characters. In the best tradition of British suspense fiction. 192pp. 5⅜ × 8½. 24720-1 Pa. $5.95

THE WIT AND HUMOR OF OSCAR WILDE, Alvin Redman (ed.). More than 1,000 ripostes, paradoxes, wisecracks: Work is the curse of the drinking classes; I can resist everything except temptation; etc. 258pp. 5⅜ × 8½.   20602-5 Pa. $4.95

SHAKESPEARE LEXICON AND QUOTATION DICTIONARY, Alexander Schmidt. Full definitions, locations, shades of meaning in every word in plays and poems. More than 50,000 exact quotations. 1,485pp. 6½ × 9¼. 2-vol. set.
Vol. 1: 22726-X Pa. $15.95
Vol. 2: 22727-8 Pa. $15.95

SELECTED POEMS, Emily Dickinson. Over 100 best-known, best-loved poems by one of America's foremost poets, reprinted from authoritative early editions. No comparable edition at this price. Index of first lines. 64pp. 5³⁄₁₆ × 8¼.
26466-1 Pa. $1.00

CELEBRATED CASES OF JUDGE DEE (DEE GOONG AN), translated by Robert van Gulik. Authentic 18th-century Chinese detective novel; Dee and associates solve three interlocked cases. Led to van Gulik's own stories with same characters. Extensive introduction. 9 illustrations. 237pp. 5⅜ × 8½.
23337-5 Pa. $5.95

THE MALLEUS MALEFICARUM OF KRAMER AND SPRENGER, translated by Montague Summers. Full text of most important witchhunter's "bible," used by both Catholics and Protestants. 278pp. 6⅝ × 10.   22802-9 Pa. $10.95

SPANISH STORIES/CUENTOS ESPAÑOLES: A Dual-Language Book, Angel Flores (ed.). Unique format offers 13 great stories in Spanish by Cervantes, Borges, others. Faithful English translations on facing pages. 352pp. 5⅜ × 8½.
25399-6 Pa. $8.95

THE CHICAGO WORLD'S FAIR OF 1893: A Photographic Record, Stanley Appelbaum (ed.). 128 rare photos show 200 buildings, Beaux-Arts architecture, Midway, original Ferris Wheel, Edison's kinetoscope, more. Architectural emphasis; full text. 116pp. 8¼ × 11.   23990-X Pa. $9.95

OLD QUEENS, N.Y., IN EARLY PHOTOGRAPHS, Vincent F. Seyfried and William Asadorian. Over 160 rare photographs of Maspeth, Jamaica, Jackson Heights, and other areas. Vintage views of DeWitt Clinton mansion, 1939 World's Fair and more. Captions. 192pp. 8⅞ × 11.   26358-4 Pa. $12.95

CAPTURED BY THE INDIANS: 15 Firsthand Accounts, 1750–1870, Frederick Drimmer. Astounding true historical accounts of grisly torture, bloody conflicts, relentless pursuits, miraculous escapes and more, by people who lived to tell the tale. 384pp. 5⅜ × 8½.   24901-8 Pa. $7.95

THE WORLD'S GREAT SPEECHES, Lewis Copeland and Lawrence W. Lamm (eds.). Vast collection of 278 speeches of Greeks to 1970. Powerful and effective models; unique look at history. 842pp. 5⅜ × 8½.   20468-5 Pa. $13.95

THE BOOK OF THE SWORD, Sir Richard F. Burton. Great Victorian scholar/adventurer's eloquent, erudite history of the "queen of weapons"—from prehistory to early Roman Empire. Evolution and development of early swords, variations (sabre, broadsword, cutlass, scimitar, etc.), much more. 336pp. 6⅛ × 9¼. 25434-8 Pa. $8.95

AUTOBIOGRAPHY: The Story of My Experiments with Truth, Mohandas K. Gandhi. Boyhood, legal studies, purification, the growth of the Satyagraha (nonviolent protest) movement. Critical, inspiring work of the man responsible for the freedom of India. 480pp. 5⅜ × 8½. (USO) 24593-4 Pa. $7.95

CELTIC MYTHS AND LEGENDS, T. W. Rolleston. Masterful retelling of Irish and Welsh stories and tales. Cuchulain, King Arthur, Deirdre, the Grail, many more. First paperback edition. 58 full-page illustrations. 512pp. 5⅜ × 8½. 26507-2 Pa. $9.95

THE PRINCIPLES OF PSYCHOLOGY, William James. Famous long course complete, unabridged. Stream of thought, time perception, memory, experimental methods; great work decades ahead of its time. 94 figures. 1,391pp. 5⅜×8½. 2-vol. set. Vol. I: 20381-6 Pa. $12.95 Vol. II: 20382-4 Pa. $12.95

THE WORLD AS WILL AND REPRESENTATION, Arthur Schopenhauer. Definitive English translation of Schopenhauer's life work, correcting more than 1,000 errors, omissions in earlier translations. Translated by E. F. J. Payne. Total of 1,269pp. 5⅜ × 8½. 2-vol. set. Vol. 1: 21761-2 Pa. $10.95 Vol. 2: 21762-0 Pa. $11.95

MAGIC AND MYSTERY IN TIBET, Madame Alexandra David-Neel. Experiences among lamas, magicians, sages, sorcerers, Bonpa wizards. A true psychic discovery. 32 illustrations. 321pp. 5⅜ × 8½. (USO) 22682-4 Pa. $8.95

THE EGYPTIAN BOOK OF THE DEAD, E. A. Wallis Budge. Complete reproduction of Ani's papyrus, finest ever found. Full hieroglyphic text, interlinear transliteration, word-for-word translation, smooth translation. 533pp. 6½ × 9¼. 21866-X Pa. $9.95

MATHEMATICS FOR THE NONMATHEMATICIAN, Morris Kline. Detailed, college-level treatment of mathematics in cultural and historical context, with numerous exercises. Recommended Reading Lists. Tables. Numerous figures. 641pp. 5⅜ × 8½. 24823-2 Pa. $11.95

THEORY OF WING SECTIONS: Including a Summary of Airfoil Data, Ira H. Abbott and A. E. von Doenhoff. Concise compilation of subsonic aerodynamic characteristics of NACA wing sections, plus description of theory. 350pp. of tables. 693pp. 5⅜ × 8½. 60586-8 Pa. $13.95

THE RIME OF THE ANCIENT MARINER, Gustave Doré, S. T. Coleridge. Doré's finest work; 34 plates capture moods, subtleties of poem. Flawless full-size reproductions printed on facing pages with authoritative text of poem. "Beautiful. Simply beautiful."—*Publisher's Weekly*. 77pp. 9¼ × 12. 22305-1 Pa. $5.95

NORTH AMERICAN INDIAN DESIGNS FOR ARTISTS AND CRAFTS-PEOPLE, Eva Wilson. Over 360 authentic copyright-free designs adapted from Navajo blankets, Hopi pottery, Sioux buffalo hides, more. Geometrics, symbolic figures, plant and animal motifs, etc. 128pp. 8⅜ × 11. (EUK) 25341-4 Pa. $7.95

SCULPTURE: Principles and Practice, Louis Slobodkin. Step-by-step approach to clay, plaster, metals, stone; classical and modern. 253 drawings, photos. 255pp. 8⅛ × 11. 22960-2 Pa. $9.95

THE INFLUENCE OF SEA POWER UPON HISTORY, 1660–1783, A. T. Mahan. Influential classic of naval history and tactics still used as text in war colleges. First paperback edition. 4 maps. 24 battle plans. 640pp. 5⅜ × 8½.
25509-3 Pa. $12.95

THE STORY OF THE TITANIC AS TOLD BY ITS SURVIVORS, Jack Winocour (ed.). What it was really like. Panic, despair, shocking inefficiency, and a little heroism. More thrilling than any fictional account. 26 illustrations. 320pp. 5⅜ × 8½.
20610-6 Pa. $7.95

FAIRY AND FOLK TALES OF THE IRISH PEASANTRY, William Butler Yeats (ed.). Treasury of 64 tales from the twilight world of Celtic myth and legend: "The Soul Cages," "The Kildare Pooka," "King O'Toole and his Goose," many more. Introduction and Notes by W. B. Yeats. 352pp. 5⅜ × 8½.
26941-8 Pa. $7.95

BUDDHIST MAHAYANA TEXTS, E. B. Cowell and Others (eds.). Superb, accurate translations of basic documents in Mahayana Buddhism, highly important in history of religions. The Buddha-karita of Asvaghosha, Larger Sukhavativyuha, more. 448pp. 5⅜ × 8½. ,
25552-2 Pa. $9.95

ONE TWO THREE . . . INFINITY: Facts and Speculations of Science, George Gamow. Great physicist's fascinating, readable overview of contemporary science: number theory, relativity, fourth dimension, entropy, genes, atomic structure, much more. 128 illustrations. Index. 352pp. 5⅜ × 8½.
25664-2 Pa. $8.95

ENGINEERING IN HISTORY, Richard Shelton Kirby, et al. Broad, nontechnical survey of history's major technological advances. Birth of Greek science, industrial revolution, electricity and applied science, 20th-century automation, much more. 181 illustrations. ". . . excellent . . ."—Isis. Bibliography. vii + 530pp. 5⅜ × 8¼.
26412-2 Pa. $14.95

Prices subject to change without notice.
Available at your book dealer or write for free catalog to Dept. GI, Dover Publications, Inc., 31 East 2nd St., Mineola, N.Y. 11501. Dover publishes more than 500 books each year on science, elementary and advanced mathematics, biology, music, art, literary history, social sciences and other areas.